Conversations About Biology

Conversations About

BIOLOGY

Edited by Howard Burton

Ideas Roadshow
INTELLIGENT. INQUISITIVE. INTERNATIONAL.

Ideas Roadshow conversations present a wealth of candid insights from some of the world's leading experts, generated through a focused yet informal setting. They are explicitly designed to give non-specialists a uniquely accessible window into frontline research and scholarship that wouldn't otherwise be encountered through standard lectures and textbooks.

Over 100 Ideas Roadshow conversations have been held since our debut in 2012, covering a wide array of topics across the arts and sciences.

All Ideas Roadshow conversations are available both as part of a collection or as an individual eBook.

See www.ideasroadshow.com for a full listing of all titles.

Edited, with preface and all introductions written by Howard Burton.

All *Ideas Roadshow Conversations* use Canadian spelling.

Contents

A MATTER OF ENERGY
BIOLOGY FROM FIRST PRINCIPLES
A CONVERSATION WITH NICK LANE

OUR HUMAN VARIABILITY
A CONVERSATION WITH STEPHEN SCHERER

SLEEP INSIGHTS
A CONVERSATION WITH MATTHEW WALKER

Textual Note

The contents of this book are based upon separate filmed conversations with Howard Burton and each of the five featured experts.

Frans de Waal is the Charles Howard Candler Professor of Psychology at Emory University and the Director of the Living Links Center at Yerkes National Primate Research Center. This conversation occurred on April 23, 2013.

Alcino J. Silva is Distinguished Professor of Neurobiology, Psychiatry and Psychology at the David Geffen School of Medicine and Director of the Integrated Center for Learning and Memory at UCLA. This conversation occurred on April 10, 2014.

Nick Lane is Professor of Evolutionary Biochemistry at University College London. This conversation occurred on October 11, 2016.

Stephen Scherer is the GlaxoSmithKline Research Chair in Genome Sciences at the Hospital for Sick Children and University of Toronto. This conversation occurred on October 1, 2014.

Matthew Walker is Professor of Neuroscience and Psychology and Founder and Director of the Center for Human Sleep Science at UC Berkeley. This conversation occurred on March 31, 2014.

Howard Burton is the creator and host of Ideas Roadshow and was Founding Executive Director of Perimeter Institute for Theoretical Physics.

Preface

Biology deeply intrigues me, both scientifically and sociologically. That it represents one of the most dynamic areas of scientific research today is so patently obvious to even the most casual observer that it teeters dangerously close to a truism, and there is certainly a steady flow of fascinating developments coming from the front lines of biological research—from genetics to evolutionary biology to our rapidly burgeoning understanding of the mechanics of the human brain.

But that is not all; because unlike modern physics, say, modern biology seems continually forced to grapple with problems largely of its own making, continuously engaged in the business of overturning unsubstantiated judgements and airy declarative statements of a mere few decades ago. I'm not entirely certain why this is the case—I personally suspect that it has to do with a culture of venerating canonical views that begins by demanding its young spend untold hours memorizing thousands of relatively arbitrary names of things, preferably in Latin—but the whole scene has often struck me as something like what I imagine physics before Galileo must have been like.

When I was young, for example, I remember being assured by confident experts that, in striking contrast to other parts of the human body, the brain was a rigidly fixed biological entity whose potential was completely predetermined from birth. And then along came "neuroplasticity" a few decades later, and everything was eventually turned on its head.

Well, revolutions are always exciting, but usually not very efficient. Just imagine how much further ahead we'd be today if we hadn't unthinkingly instilled false information in previous generations.

As it happens, each one of the five biology conversations in this collection has a similar sort of story associated with it: for a long time conventional wisdom dictated something or other that virtually everyone unthinkingly went along with, or nobody had bothered to appreciate the importance of some deeply prevalent phenomenon or shared characteristics in plain sight, and only now are we finally beginning to question our previous worldviews—often with a corresponding scientific impact that is no less than truly spectacular.

Jay Gargus describes how we've only somewhat recently begun to seriously appreciate that the roughly 99% of our DNA that is non-coding might be worth seriously paying attention to.

> *"The exome is that 1-2% of the DNA that codes for a gene that makes a protein and is associated with an aspect of the mutation process that we have a good understanding of. The other 98% of the DNA, we know that there can be mutations in those areas that do make disease, but we're really on shaky ground in terms of interpreting them. It used to be called 'junk DNA' and we certainly know now that that's not the way to look at it. It's definitely not "accidentally" being somehow carried along for evolutionary timescales..."*

Alcino Silva enthuses how his carefully constructed experiments flatly overturned the prevailing dogma that cognitive deficits attributable to a developmental disorder cannot, *in principle*, be reversed.

> *"We had these animals with a specific deficit, together with a drug that was targeted at the biochemistry of a certain gene that we believed was responsible for mediating those deficits. And the results were unbelievable. We ran this experiment enough times that we became convinced that for neurofibromatosis type one—NF1—we **can** intervene in the adult and essentially reverse the learning and memory deficits that this animal showed."*

Nick Lane describes how for far too long evolutionary biologists have paid insufficient attention to the fact that all cells we know of operate on identical energetic principles of ions and membranes.

"What does this suggest? I don't think we know, and this is not the kind of question that many people worry themselves about. It could suggest that there's a fundamental reason underpinning it that it had to be this way and other ways either don't work as well or simply never got a hold in the first place. It's different to something like DNA, where as soon as you see the structure, you understand how it works. And with this, it's just not chemistry. Everybody thought in terms of chemistry, but nobody even dreamed that the intermediate would not be just another chemical molecule, but would be a membrane with an ionic gradient across it. What I'd like to think, and what I think is the case, is that it's fundamental: that not only does it structure the way that life works here but that it alludes to some general principle that can in some ways explain that whole continuum."

Matthew Walker relates how the scientific appreciation of sleep has gone from effectively zero to a virtual panacea in a few decades.

"Sleep science has often been considered almost a charlatan science. It's only been in the past 30 or 40 years that it's become a valid scientific field of study. Twenty years ago, the way the question was posed was, 'What is the function of sleep? What does sleep do?' Now, the question has essentially been turned on its head: 'Is there anything that sleep does not benefit? Is there anything that doesn't go awry when you don't get sleep? And is there anything that doesn't gain an advantage when you do get sleep—seemingly every tissue in the body and every process in the brain?'"

Lastly, **Stephen Scherer** recalls his puzzlement at the long-standing biological conviction that genetic mutations could only occur at single sites on the genome or, in some very particular cases, with additions or deletions of entire chromosomes.

"Trisomy 21 is mainly associated with Down syndrome, but there are other big segments of DNA in a very small portion of the population

that are different from each other. 0.4% of the population have these big, big changes, and we've known about that for 50 years. On the one hand we've got those 3.2 million potential single-nucleotide changes we were discussing earlier that everyone is subjected to, and on the other hand we know that 0.4% of the population experience these large-scale chromosome changes.

"When I was teaching back in 2002, I kept thinking to myself, Biology favours balance. There must be a lot of other variants here. Why is it that we haven't seen them yet? Well, because we didn't have the tools to see them."

Well, not entirely. Of course technology played its role, but as Stephen himself describes, a principal reason why large-scale copy number variation was not discovered during the Human Genome Project was not so much because of the lack of appropriate technology, but rather because of the way the entire project was constructed.

"The Human Genome Project made a consensus sequence of what a human DNA would look like, based on a lot of individuals—I think there were 708 different donors. To come up with a consensus you have to merge them. It's like a grey picture of what a human genome would look like. And to do that, because there were lots of different pieces of DNA coming together from different individuals, you take the easiest explanation: you essentially force them together and come up with the most common, linear sequence. Just based on the design of how they went about doing things, then, you would not see these large pieces of DNA missing, because you erase that variation when you merge things."

That today's biologists are rapidly changing our world to our untold benefit is undeniable.

But if we paid a little more attention to the mistakes that have been made and a little less in forcing today's teenagers to memorize things, we could surely make the future much brighter much sooner.

Autism: A Genetic Perspective

A conversation with Jay Gargus

Introduction

Machine Repair

In his essay *Descartes' Myth*, British philosopher Gilbert Ryle coined the deliberately disparaging phrase "ghost in the machine" to ridicule any dualistic treatment that tried to formally separate our minds from the physical world. To Ryle, any claim that mental states are not the direct product of physical states is *"entirely false, and false not in detail but in principle"*.

These days, few serious scientists would seriously dispute Ryle's argument, and surely no one in the field of genetic or biomedical research.

But if the ghost has been well and truly exorcised from our corporal machines, what can we say of the machines themselves? How do they run? What causes them to go awry? And—perhaps most importantly—how might we be able to put them on the rails again when they do?

Jay Gargus, Professor of Physiology, Biophysics and Pediatrics at the University of California, Irvine, has long been occupied with these very questions. His dual background as both a geneticist and biophysicist (in addition to a medical doctor) allowed him to hew out a somewhat unusual trajectory of using genetic approaches to search for underlying physical mechanisms.

These days, he finds himself at the forefront of autism research as the Director of UCI's Center for Autism Research and Translation. Jay's journey into the world of autism and other psychophysiological disorders was arrived at through an unexpected diversion into migraine research, and is a revealing case in point of his overriding mechanistic outlook that traces back to his graduate student days at Yale.

> "*I was supposed to find mutations in a protein called sodium-po-tassium-ATPase, but I couldn't. It's an ion pump that sets up gradients across all the cells in our body to keep potassium inside of cells and keep sodium outside. If you want to trap a molecule inside of a bag, it will swell and burst. Plant cells have rigid cell walls to prevent this, but animal cells developed a different structure, so that, if they were going to keep DNA and proteins within the cell membrane, they were going to have to pick something to keep outside and that 'something' turned out to be sodium.*

> "*They then pump sodium out, which offsets the swelling so that the cell doesn't pop, and enables cell motility and many other things, but it also allows for electrical excitability, because the difference between the sodium ions outside and the potassium ions inside establishes a potential difference, which allows the passage of an electrical, ionic current. That's how nerves work, that's how muscles work, that's how all of our sensory transduction works: we know about the outside world through those kinds of gradients.*

> "*Now, I was supposed to find mutations in that mechanism. I found a whole bunch of related things that wound up being interesting for ion transport mechanisms, but I couldn't change that. So I concluded that this mechanism is so fundamental that you probably can't change it—if you change it, then you're dead.*"

But unexpectedly, years later, a new medical discovery emerges: a mutation in this very protein that is found to be associated with a rare condition called familial hemiplegic migraine. Jay naturally found himself intrigued, and then quickly flummoxed.

> "*Suddenly, someone finds a mutation in hemiplegic migraine that hits my protein that, twenty-five years ago I was supposed to be finding mutants in; and they didn't bother to look at why the protein was different. They didn't look at the function. They just said, 'Here's the mutation. This is what causes hemiplegic migraine.'*

"So, I put together a group of people, and together we were able to dissect mechanistically what happened, how the machine broke down."

The model that Jay and his colleagues put together combined a detailed understanding of the genes responsible for hemiplegic migraine, together with a detailed awareness of the particularities of specific ion transport mechanisms, such as the so-called calcium and sodium channels.

From there, a wealth of connections to other conditions suddenly seemed possible.

"Over the years, it became very clear that there were a number of sodium channels, calcium channels and these kinds of mechanisms associated with what we call 'calcium homeostasis' involved in a wide number of conditions that ranged from seizures to migraines to some of the neuropsychiatric diseases. They seemed to share a family of pathways, if you will, that we didn't really completely understand, but we got a sense of what they might be."

As Jay turned his attention increasingly towards autism, while investigating the precise role that calcium homeostasis might play as an underlying mechanism for a whole host of disorders, another medical development suddenly opened up an entirely new prospective avenue for treatment.

"A paradigm that was really important for me, which might seem at first completely unrelated, has to do with the new discovery that happened in cystic fibrosis. Just like with autism, there was no animal model of cystic fibrosis. Now, cystic fibrosis involves just one gene—I don't mean to say that the complexity of cystic fibrosis is the same as autism—but if you change that gene in a mouse you could see some things, but you wouldn't get the clinical disease that affected humans, so you couldn't make a clinical trial based on animal model data.

"So they went back and said, 'We're just going to show you that we can cure a specific, broken molecule in humans; we're going to cure that broken molecule. We're not going to give you an animal model,

> *we're going to go right in and fix that human cell; and we're going to take one mutation that we really understand very well.'*

> *"There are two thousand mutations that lead to cystic fibrosis, but they decided to pick one—a rare one, not the common one that leads to cystic fibrosis—to give them a molecule that they could screen. They were able to get a molecule that fixed the effect of that mutation; and then they were able to take that to clinical trials. Now, that's a paradigm shift, because for many years people struggled trying to get an animal model, but suddenly you could go to clinical trials without that, by dealing directly with the relevant lesion in a human."*

Autism's genetic complexity presents even greater challenges than cystic fibrosis, but that is all the more reason to carefully index and analyze all steps in the process so as to sift out so-called "biomarkers" that will help point towards a concrete mechanistic understanding.

> *"We want to understand functionality at the level of single cells; functionality at the level of single neurons; functionality at the level of how the neurons talk to one another—not for 'everybody with autism', but for each kid, each individual kid whose genomic sequence you know. We want to see all those steps, how all of these functions are tied together all the way up to EEGs and behavioural tests. We need that kind of annotation in order to be able to understand what's really going on.*

> *"What that's going to do is perhaps let us separate out different types of autism, just like the way we do breast cancer: some people are breast cancer HER2/neu positive, some will be breast cancer HER2/neu negative. I'm just using these as examples of words, but the key point is that they're markers, they're biomarkers; and that's just something we don't yet have in autism, we don't have any objective biomarkers."*

An intriguing implication of Jay's approach is that, notwithstanding its history, autism naturally becomes much more than a complex neurological condition to be treated by neuroscientists.

"Most of the arguments that have been made in model systems are at the level of doing brain slices and really looking in-depth at the brain and how it behaves. But I believe that there's a real value in looking at the simpler, core level before you get to the complexity of the brain.

"We don't have to find all of the genes, we have to find an actionable gene, an actionable pathway; and that's what I think the calcium story is pointing us towards. It's also important in that it's quite similar to the mechanisms that are involved in cystic fibrosis. It sounds like they're a million miles apart: one's in the epithelia and one's in the nerves, but it's also important to recognize that autism isn't only a neurological condition. The kids have a multi-system disease."

More evidence, then, if any more were needed, for unbridled physicalism: that the mind and the body make up one large, highly complex, inextricably connected machine, without a ghost in sight.

And if we want to keep it running smoothly, we'd best find out how it works at the most fundamental level possible.

The Conversation

I. The Lure of Genetics

Where the action was—and still is

HB: You have an MD and PhD from Yale. Were you one of these people who wanted to be a doctor ever since you were a little kid? How did you get into genetics research?

JG: Well, no, it's certainly not the case that I always wanted to be a doctor. It happened that genetics research was at a very exciting time when I was an undergraduate at Case Western Reserve. They had some very good programs there that let you do research during the month of January—they were called "414" courses.

As a result, I got introduced to some very interesting paradigms that were just evolving in bacterial genetics, and it became very clear to me that that was the way you could answer many complicated problems.

One of my faculty advisers told me, *"You know, you really should do an MD-PhD, because this stuff is moving into humans."* I then had the opportunity to go to Cold Spring Harbor Laboratory—Jim Watson had just moved there at that time—and they had training programs for postdocs, but I was able to get into it as an undergraduate, when they were just starting to do genetics in humans.

Then I went to Yale, where they were really doing some pioneering stuff, beginning to do the kinds of genetics in human cells that we knew how to do in bacteria. I was exposed to a lot of pioneering people who were really trying to understand what human mutations were. We didn't really know, at that time, whether or not human mutations were essentially the same as bacterial mutations: they did a lot of funny things that didn't exactly correlate with what we expected them to do.

It was a very interesting time; and it's just been absolutely amazing to see how rapidly human genetics has been transformed. When I was at Cold Spring Harbor, they were passing on the techniques for what we now call "banding" chromosomes, identifying every chromosome separately.

It's amazing when you think about it: you only have to go back to 1956 to a time when we didn't even have the right number of chromosomes for humans.

HB: Oh, really?

JG: Yes, we thought there were 48 back then. In 1956, they found out that we really have 46, and it wasn't until the time around when I was at Cold Spring Harbor that people were able to do banding and recognize each one. They used to be able to recognize them by their size-classes, but with banding they finally became able to actually see each chromosome separately.

Then, while I was at Yale, the first genome was sequenced by Sherman Weissman's group. It took two floors of the building and about 25 postdocs, and what they sequenced was only 3000 base pairs. It's the kind of thing that a first-year graduate student can now do in the first day of the first week of her program. It's really unbelievable how the technology has enormously driven this field over a very short period of time.

So, it's really been mind-boggling to see that. When I was at Yale as a graduate student, to move a gene, we moved a whole chromosome—and that was considered a major achievement to be able to do that. Now, we can go in and edit every mutation and we can understand all of the sequence.

It's going to take time to digest all of that information, but it's all there now.

HB: So, you were at the right place at the right time, it seems—the right place meaning "Planet Earth" essentially. Before I go a bit further, I'd like to return briefly to a throwaway comment you made earlier about how this program at Cold Spring Harbor was usually for post-docs, but you got in as an undergraduate. How did that work?

JG: I'm not sure exactly. I was at Case Western Reserve as an under-graduate, and I had done some interesting undergraduate research there, so they put me into a Master's program and I got my Master's degree at the same time as I got my undergraduate degree. One of the faculty at Case Western was going to be teaching at Cold Spring Harbor, and I just got accepted into the program.

HB: OK, so you're a very self-effacing guy, it seems.

There was something you mentioned in a video of a public lecture of yours that I saw recently, where you explicitly mention how all of us have some genetic susceptibility to diseases that are around us.

We're going to talk about autism later on, but first, I'd like to talk more generally about this idea of emphasizing the importance of looking at the world through "genetic eyes", as it were, of being able to get a deeper understanding of how we're all potentially susceptible to a wide spectrum of different conditions.

JG: Well, everybody's got the same set of genes—genes are just the names of places on chromosomes—and what we view as the "differ-ences" between people are, in genetics, what we call an "allele", or a copy: tiny changes in the sequence make for a different DNA sequence and that often makes for a different function—the best understood functional changes are understand in that kind of a manner. The point is (and this has been done for a really long time) that at a mathematical level, we can essentially figure out what the "disease load" is that people typically carry.

I do medical genetics as well, and I often talk to parents whose child has a so-called "recessive" disease, explaining to them the genetics behind it: that each parent is a "carrier", because both the mom and dad carry abnormal copies and the condition only mani-fests itself in the child when both abnormal copies are passed on.

As we explain it, we know that everybody carries mutations for probably 7–10 separate lethal conditions, which they're protected from by the fact that they get a full set of the genome from mom and a full set of the genome from dad. In other words, there's nothing unusual or "bad" about being a carrier.

In fact, what we really understand is that the best "good" for humans—and all species for that matter—is the breadth of diversity. Your genetic diversity is what lets you get through bottlenecks, through difficult times.

There is no such thing as a "good" genome, it's just that, under certain circumstances, some convey a certain selective advantage. We see specific examples of this in things like sickle-cell anemia, where if you carry one copy of the sickle-cell anemia gene, you're protected from things like malaria; but if you get two copies, then you have a serious disease that can be lethal or at least very debilitating.

That's just one of many examples of a "bad" genome that, in certain circumstances, can be very beneficial.

Questions for Discussion:

1. To what extent has our increased understanding of genetics redefined what we mean by "disease"? Readers interested in this topic are referred to Chapter 10 of **Our Human Variability** *with geneticist Stephen Scherer.*

2. Do you think that technological development in genomics over the next 20 years will be slower, faster or at roughly the same pace as what has happened over the past 20 years?

II. Genetics 101

A basic overview

HB: Let's start at the very beginning with some definitions, because if you're some guy on the street, you've probably heard terms like DNA, chromosome, genome or what have you, but it's useful to start with some tangible specifics.

In particular, at least for me, there's often a bit of a disconnect between looking at a mutation in general terms as "some structural change to the genetic structure" and the physical mechanisms in terms of the functionality of separate parts that are involved. So it would be helpful to have a bigger picture perspective on all of that.

JG: Well, I'm not going to go back into deep history, but we clearly understand that chromosomes are structures—gigantic strands of DNA—and the "action" part is the DNA part. You can think of DNA as being like "Pop It" beads—a long, linear molecule of letters, where only four letters are used to spell out all the information.

We know that there are only four letters, but the chromosomes in man, as well as in other higher organisms (it's not true in bacteria) are complicated, with lots of proteins wrapped around them that pick the times that those different pieces of DNA get exposed and turned on. That's really getting into the level of gene expression, which is an evolving field; and we certainly don't have a comprehensive understanding of how that works.

HB: Okay, let's back up for a moment. These proteins you just mentioned are some complicated, molecular things that are somehow wrapped around the DNA...

JG: They transform this long, linear DNA molecule into a much more compact structure. There are some simple, structural proteins like the histone proteins that do that, but there are also some very fancy, regulatory proteins that do more elaborate things in terms of unravelling the DNA.

But at the level of the chromosome, we know that a chromosome is a complex of a string of DNA and a whole bunch of proteins that are associated with it. And the part we understand really well is the DNA sequence, because now we can "read" that sequence.

So, at the current level, without dwelling on all the ways we've known the word "gene" in the past—which is a very interesting point, because as I told you earlier, there was still a lot of uncertainty, even when I was a graduate student, as to whether the kinds of things that happened in man were really going to be the same as the kinds of things we understood happening on chromosomes in bacteria—we can "read" the DNA now; and we routinely do read all of it.

These days, all the kids coming to our Center are getting whole genome sequencing. Again, the rate of progress of all of this is just mind-boggling. The first individual genome that was explicitly sequenced—Craig Venter's sequence—took ten years, hundreds— if not thousands—of research laboratories, and cost three billion dollars. The second genome was Jim Watson's, which probably took about one tenth of that amount of money (a couple hundred million dollars, probably) and happened much faster. Finally, the third genome was only fifty thousand dollars. Now, we're routinely getting whole genome sequences for under a thousand dollars.

Once again technology has changed matters dramatically: how well we can read the piece of DNA, how well you can read those "Pop It" beads and the letters on them. Now, we still don't know what they mean, but we know that we are able to create a file of those three billion—in fact, it's really six billion because you have to read both the copy from mom and the copy from dad, both strings of DNA, in order to get all of this information.

However, the speed and the price for doing this has fallen dramatically. So now, we're at the point where a major question is how we're going to handle keeping all this information.

When we get the information back from the sequencer, they have to mail us a hard disk, because you can't do it in any easy way right now: we have to have supercomputers just to keep the files of the DNA sequences, together with very elaborate what we call "pipelines" in order to manage that data.

Conventionally, we took the chromosome and said, "*On the chromosome, there are a lot of genes.*" Now that's still true, we still have things that are conventionally called "genes", and the things that we understand in the simple sense as genes make up about 1-2% of the entire DNA sequence.

That means that, at our current level of understanding, about 98% of the DNA on our chromosomes are relatively unannotated; and, to a large extent, we will often ignore that part of the DNA.

As the technology has changed so rapidly, we don't see that there's any efficiency in only capturing what's called the exome. The exome is that 1-2% of the DNA that codes for a gene that makes a protein and is associated with an aspect of the mutation process that we have a good understanding of.

However, for this other 98% of the DNA, we know that there can be mutations in those areas that do make disease, but we're really on shaky ground in terms of interpreting them.

Traditionally, the gene—and this goes back to the era of bacterial genetics—is a linear sequence of DNA that you read three bases at a time, and each of those three bases represents an amino acid.

Amino acids, meanwhile, come in 20 different "flavours", which are combined to form a protein. And the chemistry of that linear string of amino acids, what we call the "primary sequence" of the protein, will dictate how that protein folds up. When the chain folds up, it will make little grooves, which wind up being catalytic sites for proteins. That's a simple description, but, in essence, that's what we believe is going on.

If you make a change in one of those amino acids of this primary sequence, the protein won't fold up in the same way, which means that the site won't work in the same way and you'll have effectively broken that enzyme. And that is how we understand disease processes.

Many of our simple, Mendelian ideas (Gregor Mendel is regarded as the "father" of genetics for identifying hereditary traits—"genes"— which segregated independently of one another) we came to develop a fuller understanding of through the process of molecular biology and molecular genetics.

Starting first with bacteria, and then later back-translated to man, we developed this idea that one gene encodes a specific protein, which, in turn, has a specific function. And these functions would then, in turn, develop into networks and pathways.

For example, you might start out with a simple sugar molecule, and by changing one bond after another you could ultimately make this complicated transformation that could build something like vitamin C, or some other complicated molecule.

So in the end you might have a very, very long pathway that resulted from many genes interacting together that could do a very complicated thing.

HB: I'd like to talk about that in more detail. But before I do, I'd like to back up just a little bit. So I've got all this DNA, separated out into 46 chromosomes, which I inherit from my parents.

JG: That's right: 23 from mom and 23 from dad.

HB: And all of that represents all of my DNA that's spread out in these chromosomes. So when someone talks about my "genome", they're looking at all of that DNA, all over the place, is that correct?

JG: That's correct. Your genome is all of your DNA sequence: six billion base reads, basically six billion digits in a row.

And, in reality, when we do that, you have to imagine just typing them out (that's basically how it works). And, as a result, just like when anything is typed out, there will be errors introduced. How do we know that there's an error? Well, typically, when we're doing genomic sequencing, we want to read every sequence at least 30 times, so we read at a depth of 30-fold.

In other words, we read little pieces of the sequence and then overlap all of these "sequence reads" so that we have a huge amount of redundancy to see what the real sequence is, which is critical for being able to be confident that you have the right data set.

That's why what we've decided to do in our Center is that, even though we know that right now we don't really know what to do with that 98% of the non-coding DNA, it's still cost-effective, logical and appropriate to collect all that data at once, because a person's genome never changes. From the moment of conception, the DNA in that person is the same in all the different tissues of her body, and it never changes over time.

So, we figure, "*Why not capture it all now, at the same time?*"

All the regulatory issues are huge hurdles to overcome—to find people, to get their consent, all of that—we want to get all of the data, even though we know first what to do with that 1-2% percent that we call the exome.

To summarize, then, the genome is everything, the exome is that 1-2% that encodes segments of DNA that know how to make a protein, and proteins are the things that we understand carry out metabolic pathways. Meanwhile, we know that this other DNA really is doing stuff, and that it probably regulates matters in some complicated way, resulting in a red blood cell being different from a skin cell, or heart cell or brain cell—because those same sets of genes get turned on in different ways in different tissues. Again, that's really an evolving area.

HB: So, I want to get back to that, but just before I do, my understanding is: within these 46 chromosomes that apportion, as it were, all these strands of DNA, you can isolate things. You can say, "*Oh, there's a mutation there,*" or "*There's something happening in this chromosome over here, as opposed to that chromosome over there.*"

JG: Yes, absolutely. It's just amazing, so many of these techniques that so many of these poor graduate students and postdocs spent all of this time learning how to do have been superseded to the point where, now, you just read the sequences; you don't really have to bother to clone them so much in order to find them.

It used to be that you had to go in and "clone out" the mutated segment in order to find it, but we certainly can do that in a very straightforward way now through modern techniques.

The fundamental thing about DNA that makes it easy to carry out a lot of the reactions that we do is that DNA is a double-stranded molecule. If you have one strand of the molecule, it will allow you to create, to synthesize, because of the process of "base pairing".

I told you before that there are four flavours of bases for DNA, and one flavour always pairs with the same partner. So, if you have a string of DNA with these bases, they're going to know how to synthesize the other strand. Another way to look at that is that, if you put down on a grid a single-strand sequence of DNA, it will find its partner.

A lot of the ways that genomic sequence information came about was through what are called "microarrays", which led to a whole variety of studies where you could put down on a little chip, little strings, or sequences, of DNA and count on the fact that you could then take a whole genome from a person, break it up into little pieces, and the piece that matched, would stick.

HB: How does that work, exactly? What's the physical mechanism that's going on in that process?

JG: Yes, hydrogen bonds basically.

HB: I see. So it's a chemistry thing.

JG: Yes, but it's also a combinatorial thing, because you have to have strings of what we call "oligonucleotides"—short strings of DNA—that are long enough so that the binding energy of having a perfect match is different from having even a single mismatch.

So, all of that gets worked out; and a number of companies have been involved in doing it in a variety of different ways. The bottom line is that they continuously got better at doing higher and higher densities so that the kinds of chips that we do now can look at two million spots on the chromosome.

Now, again, to put that into perspective, I told you that just prior to the time I went to Cold Spring Harbor, we could only count the number of chromosomes. You could find something like Down syndrome, because there was a whole extra chromosome involved.

At first, we didn't know exactly which chromosome that was but we knew it was one of the tiny ones. Once they were able to "band" the chromosomes, we knew that the third copy was of chromosome 21—hence trisomy 21.

At the highest resolution of this banding technique that we were doing just 5 to 10 years ago, if you looked at chromosomes, you could resolve 700-800 places on the chromosome.

We went from just counting the chromosomes to seeing strings of DNA 700-800 sites long. Now with the modern microarrays, which have, I'd say, completely supplanted looking at chromosomes— there's not really much reason to look at them anymore; we just take the DNA from the people and put it on the microarray—we're looking at two million places.

So, we've taken another huge jump from identifying sequences 700-800 sites long to 2 million sites. And that really opened things up to a whole new class of mutations that we weren't really aware of that we call "copy number variants", and those have wound up becoming very important.

HB: When I hear you say, *"Here we've got the DNA within this chromosome, and 1-2%, as we understand it, is responsible for being involved in the production of these proteins that are related to diseases,"* I find myself a bit confused.

Now, I don't claim to know anything about this at all, but I'm thinking, *That's not very efficient evolutionarily.* If there's only 1-2% of stuff responsible for this, then there's a whole lot of needless redundancy in the system, which doesn't make much sense to me. My guess is that you guys must be missing lots and lots of stuff: there must be a lot more that's going on.

JG: Well, there certainly is; and we don't have the hubris to think that we do understand it. It used to be called "junk DNA" and we certainly know now that that's not the way to look at it. It's definitely not "accidentally" being somehow carried along for evolutionary timescales—that's pretty clear.

We believe it has a lot to do with gene regulation: how a skin cell becomes different from an eye cell and so forth. But I have to put this into perspective because you have to know what's going to be actionable. I told you about the number of genomes that have been sequenced—there are now about ten thousand, it's really been a very steep, exponential curve—it is still mind-boggling that every time you sequence a new genome now, you'll find tens of thousands of mutations. What I mean is not just tens of thousands mutations that are different from the "standard" sequence, but tens of thousands of mutations that have never been seen in anyone before.

HB: Tens of thousands?

JG: Tens of thousands of unique mutations that have never been seen before. So, you try to guess what's meaningful for the disease in that kind of noise and you don't know where to go.

That's why, for simplicity, when we're doing these studies in a clinical setting, we'll often only focus on the exome, that 1-2%—

HB: Because you understand that.

JG: Well, I'd like to back up and say that, while we know the names of the genes, we don't necessarily know what many of them do. Huge numbers are "unannotated": they just have a serial number—we don't know what their role is, they're just a name.

But even within those genes that we understand, where we know their protein-coding segments, you find this same problem. It's not at the numbers of tens of thousands, but you find way more than you'd expect.

If I have a child who's severely affected, the only way we can even start to make sense out of that is to do what we now call the "trio": we do dad's sequence, mom's sequence and the child's sequence.

The only place you can get a signal that's rich enough is to look in the child for a sequence that neither mom or dad had, so you're looking for a new mutation, because that is the most promising way forward to understand what is going on.

But even if you do that you still have too many, so you have to do a lot of what's called "in silico" genetics—in silico, meaning that you're letting the computer crunch the numbers.

HB: So, hold on a second: you've applied a Latin name to this technology?

JG: Well, we have "in vivo" which means "in the living animal", "in vitro" which means "in tissue culture" and then "in silico" which means you're crunching the information with technology.

HB: Sure; I'm OK with Latin expressions per se, but that sounds pretty funny to me. I could imagine Seneca, for example, saying, "in vivo," but not "in silico".

JG: Well, in any case, you try to use all this annotation, because you're still going to have more new mutations in each child that you do than you're going to be able to figure out.

You have to say to the computer, "*Tell me what mutations are really damaging, those that really break how those proteins fold*," and then it will sort through and find the ones that could really damage the protein.

Then you'll promote those as being the most likely to explain what happened here. It's a very big data set that you're dealing with; and so, in order to understand something—because you know you're not going to be able to understand everything—you need to simplify things.

That's how we enrich the probability of finding something; and that's actually become quite efficient. Now, if we do a trio in a child who has a very strong, clinical phenotype, we can sort out even a fairly exotic presentation—where even skilled geneticists can't tell what the child has—about a third to a half of the time.

That's not a bad batting average; and the costs have come down to the point where going to the sequence early is the most logical thing to do.

Already, all states do newborn screening programs. The kids get a heel stick, which is put on filter paper and mailed to the state lab. They now use a technique called "tandem mass spectrometry", where they're really looking at metabolites—none of the state labs are looking at the DNA yet—and they'll look for things that cause these rare diseases that are profoundly damaging but can have an intervention done to let the baby have a more normal life.

The classic example is PKU, phenylketonuria: where if you feed the child a normal diet, they'll become profoundly disabled and have a severe intellectual disability. But if you just put them on a special formula, they lead completely normal lives.

When interns look at some of these kids we're seeing in our clinic, they don't even understand why we're seeing them, because, functionally, they're completely normal. So there are a lot of examples where you can really—with a very specific intervention when you understand the mutation—fix it; and so conditions like PKU led to this idea of screening all newborn kids.

In California, we went through this transition less than ten years ago. Ten years ago, California looked at only four diseases: if you were born with one of those four diseases, it would be picked up.

Geneticists eventually convinced the state to move to this new technology of tandem mass spectrometry that now looks for some 30-40 diseases, and all the states pretty much do it now in the same kind of way. They'll do that unhesitatingly, because it's so unbelievably expensive and so damaging to society to take care of an individual who's been incapacitated needlessly.

HB: And obviously a huge moral issue too.

JG: Exactly. It's both benefiting their constituents as well as being very cost-effective. The cost of the test is getting close to the point when you might just want to start reading the DNA. Many of the machines that do that are down in California, so why not start doing it here?

HB: And this comes back to what you were saying earlier of the costs coming down so much, from billions of dollars to just around one thousand dollars.

JG: Yes. And if you scaled-up to the point where you were literally doing every kid born in the state of California, the price would probably come down even more dramatically. Then the question becomes, when do you make that switch? Ultimately, we are going to be capturing everybody's DNA.

HB: It's a market decision, presumably. At some point it makes sense economically.

JG: And you're probably also going to run into issues with civil liberties—people asking, "*Do you really want to have the state of California holding your DNA sequence?*"

HB: Well, personally, I couldn't care less: anybody can have my DNA if they want it.

I'd like to return to the question of these mutations themselves.

There's this understanding that something goes wrong—the sequence isn't replicated in the way one would normally expect it to be replicated—but we haven't talked about factors that might actually cause that.

As I understand it, there are several different classes of factors. One can do something very nasty like, say, bombarding people with ultraviolet radiation or something like that—presumably that might somehow change this sequence or cause something. So, there's some sense of external, environmental issues, but might there be other factors that cause these DNA mutations?

JG: Well, there are certainly a whole host of environmental factors: there are chemical mutagens, there are radiation-types of mutagens; and we do understand how they happen.

Everything in chemistry is statistical—all these bonds are flipping around all the time—these external factors perturb the system so that you increase the chance of making an abnormal base-pairing, bringing in an unconventional base; and that's a mutation. Once you've paired to a rare conformation of the partner, that's now locked in: you have no ability to change that anymore.

HB: So, it really has to do with these structural bonds, then.

JG: Yes. As I said, there are four "flavours"—two big guys and two little guys. A big guy always pairs with a little guy; and there's one set of the pairs that use two hydrogen bonds and the other set uses three hydrogen bonds. Those molecules that are ready to make two or three go through fluctuations where, for rare conformations, they're ready to make three when they usually make two, or make two when they usually make three; and so they can bring in an abnormal partner.

HB: External factors might be able to somehow causally enhance this rarity, or whatever.

JG: Yes. Things like that are mutagens; and we understood mutagenesis before we really understood the molecular basis of mutagenesis—we've long had an empirical sense of the kinds of things that do cause mutations.

Those are conventional mutations. And again, the way we understand conventional mutations is really pretty simple. I told you earlier that the DNA gets read three base pairs at a time, so instead of having the three base pairs that are normally there, you change one of the letters and you now spell a new amino acid which puts a different structure on the chemistry of the protein that then makes the protein fold abnormally.

Most typically, the easiest examples to understand are when we "break the machine" so that it's a loss of function. We geneticists like to have good anchor points. If we delete a gene so that it's gone, lost from the system, that's going to involve a loss of function.

That's the typical situation for this whole host of diseases that I said are recessive diseases: where mom and dad both have a healthy copy and a damaged copy of a gene and, by chance—it's a one out of four chance—the child gets both damaged copies.

The class of diseases that are much trickier to understand—and that are really important in neuropsychiatric diseases—are the dominant diseases. That's where the person has one perfectly good copy of the gene but one mutant copy of the gene.

In that case, we're in a situation where the paradigm is that you've possibly had what's called a "gain of function": where the mutant does something different from what the "wild-type" copy does.

We know from these recessive diseases that the way that our genome has evolved, the way it's engineered, is that, most of the time, for almost all of the proteins in our body, we have an excess of that protein.

In PKU, for example, that condition I was talking about earlier, if you have only half of the amount of that enzyme, that critical enzyme, you're completely normal: your body is working exactly the same as it's supposed to, you just have an excess of the proteins available. You really have to go down to about 5-10% of the normal level before you start seeing any deficiency.

Now, how can it be that you have some disease where you eliminate one copy of a gene and you start seeing the disease? That's tricky.

You can say it's what we call "haploinsufficiency": that is, half the dose of the normal gene is not enough to get by on—and we only have rare examples where we really understand that very well.

Or we can say that the mutant protein has taken on some kind of a gained function, some kind of a new activity. It might be a toxin activity that the native protein didn't have.

We understand recessive pretty well: we generally know how losses work. But gains are tricky, and will take a lot of work.

HB: Could it also be that, somehow, this mutant protein, as it were, "interferes" or eats the other protein, resulting in a greater diminishment than you would normally have?

JG: That's a way of saying "toxic", yes: it might tie it up in some kind of an abnormal configuration, or what have you.

Questions for Discussion:

1. In what ways do you think the rapid advances in computer power will impact our understanding of genetic structure?

2. What do you think Jay meant, exactly, when he said, "We understood mutagenesis before we really understood the molecular basis of mutagenesis"?

III. Towards Disease

Intervening in metabolic pathways

HB: I'd like to move up to a higher-level now and sketch out a structural path for how we can use genetics to understand, and to eventually prevent or cure, some of these diseases that we're talking about. Here's my understanding: please jump in at any moment and tell me when I go awry.

My sense is that there are some physical mechanisms that are happening with people who suffer from various diseases; the reason why, I, say, would suffer from a disease that others wouldn't suffer from is that there's been some mutation in my DNA that interferes with the normal production of proteins that do their thing.

And as a result of this mutation, I don't produce those proteins— or I produce mutant proteins or something—so from this functional perspective, those proteins then go ahead and do "bad" things or somehow interfere with established, mechanized structure of what should ideally be going on.

Which means that, presumably, if I can find a way to isolate what's making those proteins do what they're doing the wrong way, or if I can impede them from being produced to begin with or whatever, then I can solve the problem.

I can't solve the problem in the way of going to the original source, of course—I can't change the DNA of what is causing this to begin with—but I can identify what's actually going on, and then I can treat it from the secondary level based on an understanding of what I'm producing (or not producing). Is that a fair assessment of things?

JG: Yes. And it also makes the very important point that the reason that geneticists are interested in understanding the genetics is not to say, *"Mom and Dad, you made this problem in your child."* That's obviously of no interest to anybody.

The real reason is that the genes give us a blueprint of how this disease came about and that the better we can understand a molecular defect, the more likely it is that we can intervene; and that intervention is very unlikely to be at the level of replacing the gene, just as you said.

There certainly has been a wave of trying to do gene therapy, of replacing defective genes, but that's going to be very limited. There will be certain places where you might be able to do that. The hope had been that, if you needed to have a certain gene expressed in blood, for example, then you might be able to replace the cell in the blood and actually correct the broken gene; but for many diseases, you'd have to do it in every place in the body, which naturally becomes quite daunting.

It's not that likely, then, that we'll be able to go in and fix the gene. That means that you have to be somewhere quite close to the gene; and that might be—if you're lucky, and that's where we've been most successful—in the direct protein product of that gene.

I think that, for some of the neurological diseases, we might need to be one step further. But we can't be a thousand steps away: we have to still be anchored very close to where the genomic information is telling us we need to intervene to fix what's been broken. And there have been a variety of ways of doing that.

Up until very recently, we could make the diagnosis of some of these very specific diseases, but we couldn't do anything about it. But as the genetic revolution has moved forwards, we're now able to make the missing protein in tissue culture.

I'm the lead on a clinical trial where they actually engineered a gene into a chicken cell and from that cell grew a chicken that lays eggs from which you can purify from the egg white a protein that you give the patient to cure the disease. He's missing one protein; you give him the missing protein and that's able to fix the disease.

There are times when it's relatively simple to deliver what's missing, but there are a lot of diseases where you just can't deliver the protein product that's missing, because it has to get inside the cell.

In addition, a lot of things have to get across what we call the blood-brain barrier. The brain is a protected region: you can't put things into a vein and expect that it's going to get into the brain; that won't happen. That becomes a real trick for a lot of these diseases: how you get things into the right place where they can actually do what they need to do.

Drugs have proven to be very important but, again, the genes have been the link that lets us do that. In one sense, if you know the gene that causes the human disease, you suddenly know how to make an animal model of that human disease: you "break" the same gene in a mouse, for example, and you've pretty much got the same disease.

That doesn't always work, and I'll come back to some examples where it doesn't work, but that's been a very important reason why it's important to know the genes. If you know the genes, you can make an animal model, you can know what the metabolic pathways are, you can know where you need to try to intervene. It's our strongest argument of how diseases come about.

HB: Let's take a concrete example, then, of what you did with migraines. Talk me through what you did here, because that struck me as very illustrative of some of the processes that you're engaged with. First off, my understanding is that these familial hemiplegic migraines are not very common, as I understand it. But perhaps that maybe makes it easier, in a way.

JG: Yes, they're very rare, which is reflective of a geneticist's paradigm: we've always said that, by understanding the very rare genetic mutations that cause a disease pathophysiology that's going to illustrate for us how the common diseases work.

The study of familial hemiplegic migraines was, if you will, my "entry" into these different neuropsychological diseases. As we mentioned, they are very rare, accounting for a tiny fraction of 1% of the people who have migraines. Migraines themselves, it should be said, are very common: they're the most common neurological disease, with well over 10% of people affected by them, whereas hemiplegic migraines are very rare. Hemiplegic just means a weakening on half of the body—that's basically what happens in that disorder.

It's a very specific kind of migraine, and it's one that is highly heritable: you can follow it through families. It behaves in this unusual pattern that I mentioned to you earlier of dominant inheritance, which means that if you catch only one copy from either mom or dad then you will show symptoms of the disease; and, statistically, half of your kids will show these symptoms.

You can follow, then, these dominant mutations through families; and you can get a very strong statistical signal that there's some specific gene that's doing this.

That's allowed us to map those genes—where "map", in the old days, used to mean finding the place on the chromosome—but we don't need to do that anymore, because we can just read the DNA sequence.

At any rate, now we're able to know where that abnormality is coming from on the chromosome. Then the question is, *How did that make the disease?* Well, you often can't tell by just looking at one gene, but it turns out that this familial hemiplegic migraine had three different genes; we were involved in finding the second and third one, but they all converged on a common kind of pathway.

The first one was a calcium channel, and I'll explain what that is in a second. The second was a protein that I'd worked on as a graduate student at Yale. I was supposed to find mutations in a particular protein—called sodium-potassium ATPase—by making them in tissue culture, but I couldn't do it.

Sodium-potassium ATPase is an ion pump that sets up gradients across all the cells in our body to keep potassium inside of cells and sodium outside. It's called the sodium pump, and while we're sitting here, about one-third of all of our metabolic energy is going to feed that ion pump.

If you want to trap a molecule inside of a bag, it will swell and burst. Plant cells have rigid cell walls to prevent this, but animal cells developed a different structure, so that, if they were going to keep DNA and proteins within the cell membrane, they were going to have to pick something to keep outside—and that "something" turned out to be sodium.

Sodium is then pumped out, which offsets the swelling so that the cell doesn't pop (establishing a so-called Donnan equilibrium), which means that now you can have cell motility and many other things.

This also allows for electrical excitability, because the difference between the sodium ions outside and the potassium ions inside establishes a potential difference, which allows for the passage of an electrical, ionic current.

That's how nerves work, that's how muscles work, that's how all of our sensory transduction works: we know about the outside world through those kinds of gradients.

Now, as I said, I was supposed to find mutations in that mechanism. I found a whole bunch of related things that wound up being interesting for ion transport mechanisms, but I couldn't change the fundamental mechanism. So I concluded that this mechanism is so fundamental that you probably can't change it—if you change it, then you're dead.

We knew that one drug that could inhibit this very important pump was ouabain, which is related to digitalis, one of the drugs that people who have heart disease often take. We could stick a mouse cell and a human cell together—that's how we did gene transfer in the old days, you just stuck two cells together—and it turned out that you could kill off the human cell a thousand times easier with this drug because the mouse pump was resistant. And you can always transfer mouse chromosomes into human cells, which was a gimmick we had been using—

HB: Why were the mouse cells more resistant to this drug?

JG: I don't know. I'm not sure if anyone knows. But at any rate, it was a gimmick that we used in what's called "somatic cell genetics", ages ago when we had very crude techniques to move genes around.

So, we thought we'd be able to understand this pump but, as I said, it wound up being very hard to get mutants into it.

Then, suddenly, someone finds a mutation in hemiplegic migraine that hits my protein that, twenty-five years ago I was supposed to be finding mutants in, but they didn't bother to look at why the protein was different. They didn't look at the function. They just said, *"Here's the mutation. This is what causes hemiplegic migraine."*

So I put together a group of people, and we got involved with Rhoda Blostein and her group at McGill University who really knew how to do some very sophisticated kinetic studies, and together we were able to dissect mechanistically what happened: *how did the machine break down?*

That ended up being pretty important, because then you could see how things functioned in the same kind of pathway that the first calcium channel worked in, and that also pointed out how the third hemiplegic migraine gene—this is the sodium channel I mentioned earlier—worked with the others: how all three worked together.

They're all involved in moving ions across the cell membrane, they're all involved in this electrical activity. And that was really important. If you look in textbooks, they still talk about migraines as being a "vascular" headache, but these genes aren't even expressed in vascular cells, so clearly that's not the common denominator of what migraines are.

Even though I'm looking at a tiny fraction of 1% of all migraines, we'd better think about migraines in terms of the neurons, because that's the only place that our genes are being expressed. And even more interesting than that is the fact that those same genes had different mutations in them that were found in epilepsies, in seizures.

That argued that, mechanistically, that the "genetic architecture" responsible for pathogenesis in seizures is probably similar to the way you make pathogenesis in migraines—they share some things in common.

That also gave evidence for why, empirically—because there aren't a lot of good drugs to treat neurological conditions—a lot of the doctors who are taking care of migraines found that anti-seizure medications were useful. So it all started to make sense: it's not just randomly affecting neurons, it's affecting the same kinds of mechanisms.

All of this information started congealing, if you will, around a certain set of mechanisms for disease. Then we noticed that there were "hits" on some of these mechanisms, particularly the third mechanism—the sodium channel—that had mutations associated with autism on it. The autism mutations on that gene were similar to the migraine mutations on that gene; they were different from the mutations that were causing the seizures on that gene. So we started thinking more about how this mechanism folded into autism.

HB: And when you say, "*Similar to the autism mechanisms*," what does that mean, exactly?

JG: Well, so these are trans-membrane proteins. That means that they go across the boundary of a cell. We found that the seizure mutations tended to occur right in the bi-layer region—in the membrane region—whereas the migraine and autism mutations were in the little loops that dangled underneath the membrane.

HB: OK, but how do you even know what an autism mutation is to begin with?

JG: At the time when we did this—and again, it's not that long ago, but the world is very different now—there were some small studies done in which people had found a series of mutations in a certain gene for a small number of families associated with autism. Having these studies allowed us to carry out what are called "genome-wide association studies" where we looked for places on chromosomes associated with different diseases.

You could do the design in a variety of different ways, but essentially you'd take a number of control individuals and a number of individuals with a specific disease or condition—it could be something like autism or schizophrenia, it could be high blood pressure, it could be height, it could be a variety of these more quantitative kind of traits—and you try to find places on the chromosomes that were associated with the two different groups, trying to get an understanding of what it is, exactly, that distinguishes them.

Over the years, it became very clear that there were a number of sodium channels, calcium channels and the kinds of mechanisms associated with what we call "calcium homeostasis" involved in a wide number of conditions that ranged from seizures to migraines to some of the neuropsychiatric diseases. They seemed to share a family of pathways, if you will, that we didn't really completely understand, but we got a sense of what they might be.

HB: This is presumably a sort of statistical argument that is being invoked?

JG: Exactly—well, that's my view. I would say that it is statistical; and so, therefore, it's a candidate. It's being promoted as more likely, but as a biophysicist, I really need to know how it's doing this. I've got to see the mechanism behind it.

But often, geneticists will say, "*Look at the statistics,*" and that's it. "*It's statistically significant, so that's the cause.*" But you can be tricked. And, anyway, you can't do anything about it unless you understand how it works. So, I think it's critically important that we fully annotate these kinds of things. As I said, when the group found a mutation in the sodium pump, and they said, "*That causes the disease,*" I thought to myself, *Well, **how** does that cause the disease?*

HB: Well, as a complete outsider, I must say that this is something that I've always had a problem with—this conflation between statistical support and cause that people often make.

To take an obvious example, we've all heard that smoking causes cancer. So I ask myself, "*Well, why do we say this?*" And the answer turns out to be something like, "*Well, we have huge statistical evidence to support this idea that people who smoke are more likely to develop lung cancer."*

Of course, I'm not denying this. I'm sure that there is overwhelming statistical evidence to support this claim. But I think to myself, *What about people who have smoked for fifty years and never developed cancer? What's going on there?* To me, that's very interesting, because, while I don't doubt the statistics—I don't doubt that a correlation certainly exists—what I really want to know is the mechanism. For me, the statistics are really all about pointing me in the right direction so that I can eventually discover the underlying mechanism.

JG: Yes. Moreover, I'll add that, in the old days at least (you don't see this quite so much anymore because of the laws of smoking in public places), if you went into a psych ward, you'd see that patients who have schizophrenia all smoke. We think they're self-medicating—we don't think that smoking is causing the schizophrenia—but there's a very high correlation. It's actually intriguing that, in autism, you don't see smoking at all. So, there's something intriguing about that but I don't know what that is. Anyway, that's way out in left field.

Questions for Discussion:

1. What does Jay mean, exactly, by "pathogenesis"? How would you describe that to a friend?

2. Do you agree that exceptional cases of resistance to a disease or condition might be particularly useful to understand its underlying mechanisms?

IV. Autism

Myths, current understanding and challenges

HB: I'd like to talk in more detail about autism. You spoke of how your research with migraines led, through this calcium channel you were mentioning, to connections with other neurological conditions, and you mentioned a specific link with mutations that are associated with autism. But before we get into the work that you're doing and the future, in terms of what you hope to achieve, let's first try to clear up a few misconceptions that people have about autism.

One misconception is this notion of a particular environmental factor or a particular environmental cause—namely vaccines—which came into the news some time ago. I don't know exactly what the dates are, but there certainly was a period when there was widespread speculation that what was causing autism was linked to childhood vaccinations for a particular disease.

And my understanding is that there have been several rigorous studies where people took this idea very seriously, and those studies have shown zero correlation whatsoever between the vaccines and the development of autism. Is that correct?

JG: That is absolutely true.

Maybe I should just start with what autism is. Autism is a clinical syndrome that really only has a clinical diagnosis. It's based on having very simplified speech, repetitive behaviours and a loss of reciprocal, social interactions; there's a constellation of clinical findings.

At a research grade, it involves a very elaborate set of neuropsychiatric tests, but there's no blood test, there's no straightforward "yes or no" answer.

You can really only make the diagnosis once the child is two, as they really don't have enough in the way of clear, clinical evidence to be able to make that call earlier. You might become suspicious when witnessing prolonged delays in speech development. Most kids start using words at about one year of age. So you might worry about a child who's late in developing speech. The Pediatric Society has just come out with new guidelines on how to evaluate these sorts of things.

But while diagnostic guidelines are progressing, the situation in autism is that there's a time window in which you can make the diagnosis. It turns out that that time window was the time window when kids are getting their measles, mumps and rubella vaccines.

The timing of when you get those vaccines is basically decided by the rate at which our immune systems mature. You can't effectively immunize a baby and expect her to have a response. So, kids are getting their vaccines around 18 months, and the typical time when you start really being able to see the symptoms of autism is around that same window of time.

This brings us back to our earlier discussion when we were talking about correlations. People started making correlations between the two. Now, the "critical" publication of Andrew Wakefield's was really criminal. It was a situation where a number of lawyers were sponsoring the clinical trial, and people who were involved in the lawsuits were in the trial. He eventually got barred from practicing medicine in the UK (he lives in Texas now) and The Lancet, the journal that published the paper in 1998, has officially retracted it.

As you said earlier, many studies have gone through looking at this in a very rigorous way, and there is not a drop of evidence that the vaccines have anything to do with it. The tragedy of all this is that people have really forgotten what measles encephalitis and whooping cough look like. We now have a situation where, in relatively affluent communities, people are electing not to immunize their children, and a young infant winds up catching an infectious disease.

Now, maybe that disease won't severely damage your child that you didn't immunize; but maybe it will, maybe your child will get measles encephalitis. And that is a horrible, horrible disease. People have just forgotten what these diseases look like, and they think that the vaccinations are something that's just "threatening" their child, but they may be jeopardizing their own child with a really horrible outcome.

This is something that really bothers me. I know that there are still people who strongly feel—not scientists, I mean mom and dad out on the street—that vaccinations are a bad thing. There are even families who have elected not to vaccinate their children, saying, when some of these unvaccinated children are eventually diagnosed with autism, "*Well, at least I didn't vaccinate him.*"

HB: What do you think that's attributable to? Is it a lack of understanding of the scientific process? The fault of the media, for playing these things up? I know you don't know, but I'm just asking you to speculate.

JG: Well, I certainly don't know. There are a lot of Hollywood stars and fashion models and people like that who have kids affected with autism and speak about this. And the way our society works is that, if you're an actor, you have a lot of credibility in whatever you want to talk about, so it doesn't seem to matter.

HB: So, it's Hollywood's fault? I'm happy to blame them.

JG: Well, I just don't know, but I know that it's immutable. I know there are some folks who are just so adamant about this that they don't want to hear your arguments.

HB: Well, of course one has to be sympathetic to the perspective of the parents. When you have an autistic child, I'm sure it's very tempting to look for something or someone to blame for the circumstances. That's perfectly human.

Anyway, so we've established the lack of any scientific evidence between autism and vaccines. One other thing that I'd like to discuss before we get into a discussion of the potential molecular mechanisms of autism is this question of the prevalence of autism and whether or not it's changing.

Some talk about an "autism epidemic", while others say that that's just an epiphenomenon: that the numbers aren't actually changing at all, but what's actually happening is that we're just diagnosing people better and better, and we're including, within this notion of an autism spectrum disorder, more and more people who previously weren't regarded as autistic.

That is, some say that nothing is really changing in absolute terms, whereas others claim that things are changing drastically. What's your sense of the lay of the land?

JG: Initially, I was quite sceptical that the numbers were changing, but I think it's pretty clear that the numbers really are changing.

There is definitely much more recognition and awareness of the disease. When I was an MD-PhD student at Yale, the chief of the hospital would meet with us and take us onto the ward to see the most exotic case in house that week.

One week, we were taken to see two brothers with autism, but if I did that today, my students would think I was nuts, because they can just walk down their street and see a kid with autism.

Clearly, something has really changed. There is certainly better recognition of the disease, everybody now knows about it. You're not going to walk into a doctor's office and find that she doesn't know what autism is.

Sometimes, the way that services are provided by the state plays the role of a somewhat irrational driver as well. If a child in California carries the diagnosis of autism, for example, he is entitled to specific services. Now, that's good—those services are certainly warranted—but this sometimes interferes with the bigger picture.

So, for instance, the child might have Down syndrome—trisomy 21. While we know exactly what that is, such children are not entitled to those behavioural, developmental services unless they also carry this extra, specific diagnosis of autism—they have to have autism as well in order to be entitled to those services.

To me, that's irrational; it doesn't make a lot of sense. I'm certainly not denying that those services are absolutely needed for people who have autism, and I'm certainly not advocating that we get rid of them. My point is simply that you have a ratcheting up of the diagnosis because of the fact that you are entitled to certain services only if you carry that diagnosis.

HB: Is that only in California? Because if it's only in California, presumably one can filter it out in broader-based statistical studies.

JG: I honestly don't know how the laws are written elsewhere. But I think that, at least here, the connection of associating the diagnosis with having specific services provided winds up being a driver.

But with all that said, the numbers are just frightening, worldwide. I think something is going on that is increasing the frequency in a very real sense. I don't think it's just a broadening of the definition, I don't think it's this "ratcheting up". I don't think it's just an earlier awareness. I think that there is, at least, a significant component that is an increased frequency. I don't know what it is.

What I do know is that, even in very simple models, "environment" is infinitely complicated. What do we mean by environment? A geneticist calls the genome a mapping function for the environment.

If you know a person's genome, and you have the environment, what you wind up seeing is how those two things interact with one another. I told you about that example of PKU. PKU is absolutely a genetic disease, but if you change a specific component of the environment (namely what you feed an infant), you wind up with an outcome that's very different.

So what is it, then? An environmental disease? A genetic disease? It's how those two work together, clearly. But with all the complexity of what the environment is, we can't usually figure out what the relevant environmental factors are until we know the genetic network that it's working through.

Once again, that puts a priority on understanding the genes first, even if you believe that the environment is the important part. Now, I know that an epidemic can't be genetic—genes do not change that fast. And since we know that the genes aren't changing that fast, it must be something that's "environmental".

But I don't know, and I don't just want to start guessing. The best approach is to first understand how the genes are working and then we'll be able to find the environmental impact points. It will fall out directly from the way you start doing screens to look for drugs that can fix the problem—you will find things that exacerbate the problem.

I don't mean necessarily in people—I mean at a protein level, at a cellular level, at an animal model level. As you're setting up to look for things that make the situation better, you will start having an understanding of the things that can make things worse, so then you'll be able to unwind it. I think that's how it will get unwound.

All the examples that I'm aware of where we've really figured out how the environment works, we had to know where it is; because there are just too many things that it could be.

HB: Presumably, it comes down once again to this idea of looking for a specific mechanism and discovering what enhances that, what impedes that, and so forth.

JG: Exactly.

HB: So, what's our best guess so far in terms of what this mechanism might be?

JG: Well, from the genetics level, we know that the kinds of signals that we're seeing are often hitting these proteins that are involved in calcium signalling. Our hypothesis has very strongly gravitated towards that.

Some people might say, "*It has to do with synaptic function.*" Synapses are where cells connect with one another; and most typically, people are focusing on where neurons connect with each other.

I believe that when they're looking at that level, they're looking too far away from where the primary, genetic lesions are; they are more downstream, in the network.

HB: But that's what I wanted to ask you, because the obvious question is, *How unique is this?* I mean, if this calcium mechanism is somehow involved in every single model of every single neurological condition—if it's so incredibly widespread—that's not terribly useful to me. So, give me an argument as to why that's correlated strongly with autism.

JG: The correlation with autism is really strictly a genetic argument at this point, there's no question about that. There are some that are really strong compelling arguments from single gene hits, from asking, "*What do we find when we take a bunch of controls and a bunch of kids with autism? What genes get hit?*"

Remember how I told you how our understanding evolved: first we were able to count the chromosomes; then we could identify each chromosome; then we were able to identify about seven hundred spots on the chromosome; then we could see about two million with the microarrays; and now, finally, when we're doing whole sequencing, we're looking at all six billion spots.

Now, at the level of reading genomic sequence, signals begin to appear. At the level of looking at the microarrays, signals were not appearing.

We haven't spoken yet about this notion of common disease, common variant. Is this a fair time to mention this?

HB: A perfect time, yes.

JG: Okay. Well, first off, I should say that the microarrays weren't suddenly two million; they started off at a couple hundred and then quickly ramped up to that level, but once we got into the era of being able to run microarrays, that meant that it was possible, in a cost-effective way, to do large cohorts of controls and "affecteds" and see what markers came up.

Now, geneticists only have a signal if there's a difference. That is, if everybody has the same DNA sequence, you can't watch anything move through the generations, because it's all the same. You have to have a difference between "A" and "B", and then you can watch who gets A and who gets B. There has to be a dependable difference.

On those microarrays, then, you had to pick the pieces of DNA to put on the microarray where 5% of the people would give you one flavour and 95% of the people would give you another.

You wouldn't put a spot on that DNA where only one in a million people had a different flavour, because you might never get that particular person. So, those arrays had to be biased towards only looking for common variations, otherwise you wouldn't be able to run a big enough study.

HB: And by "common", you're looking at, say, 5%?

JG: Yes, that happened to be the kind of cut off that people used. And it turned out that nothing turned up, basically. It's not that it was a complete failure, but there were only a couple of successes. Some of the big successes were in retinal degeneration—there were a couple examples where they did find a common variant that accounted for the common disease.

At any rate, that was the argument: that the mutations that make these common diseases will be very common in the population, so we will be able to find them with these microarrays.

Now, that was a reasonable hypothesis—it was the only kind of way that you could run that test—but the damaging part was to say, *"Forget about the rare mutations that you geneticists normally know how to study and understand what they mean; those are not related to these diseases."*

HB: Because practically they couldn't actually do it.

JG: Well, they certainly couldn't do it this way. But it also said, *"Don't even bother looking at them in model organisms."* It was a very "anti rare-disease" orientation.

Well, this was the attitude. And it turned out that they could explain, using this "common disease" approach to autism, I'd say somewhere in the range of 2-5% of the genetic variation. We know that autism is a highly heritable disease: 90% or so of the impact of the disease comes through the genetic architecture, from the genes.

So there's a lot of heritability to be found. I'm not going to sort out how we define heritability, but the genes that you're able to follow that have common variants only account for a tiny sliver of that, which means that something else is causing this.

HB: As I'm listening to you say this, it seems that this "common variance" hypothesis is a very reasonable one, but one reason why it might be wrong, I would speculate, is that things are just much more complicated than we had originally thought.

It is simply not the case for a condition like autism that there is some universal, or near-universal mutation, or set of mutations, that is responsible for this condition—that, in fact, what's probably going on is that there are different mutations all over the place for all sorts of different people on the autism spectrum and it's a real mess to sort it all out.

JG: And, in fact, now that we've moved on to the level of reading genomic information, that does seem to be what's happening: there's a huge amount of heterogeneity—different people are getting hits in different disease-causing genes, and each one of the genes only contributes a small increment of the risk.

It's probably not simply additive, but likely what a geneticist would call "epistatic". That means that there are complicated interactions: that if you break one gene, you're okay, if you break the other gene, you're okay but, if you break them both, you're in trouble.

HB: So there's obviously something going on that has a law-like explanation. But maybe it's a super-complicated law-like explanation. Maybe we have an equation with a thousand different variables or something, and we need to understand not only what the variables are, but also their coefficients—how they interact and so forth. So, there's a law out there somewhere that involves these thousand variables that's incredibly complicated to try and figure out; but right now, at least, we can eliminate the possibility of having a law with only one or two variables.

JG: Yes, I think that's a very powerful way to have said it—I've not heard it said that way, but I think that's clear. We know that there are, at least many hundred, some would say even more than a thousand, genetic loci—places on the chromosome that contribute towards the risk of autism.

Now, the problem with that is that autism therefore becomes very much like cancer. That's pretty much what the genetic architecture of cancer is: that even for two people who have thyroid cancer, if you read the DNA in their tumours, they'll be quite different from one another.

So, we've really come to understand cancer as this very complicated genetic architecture. You may have some very important common drivers that push in a big way towards a certain kind of cancer—things like some of the breast cancer genes and so on—but the way that that cancer evolved has involved a whole variety of different changes.

We are starting to work this out. I don't think it's insoluble. But the question is, *How are you going to do a study when, if you take one child with autism, he really has a "different disease" than another child with autism?*

It's this heterogeneity, all these factors with these different coefficients that you're talking about, that results in one person being put together quite differently from somebody else.

Our approach in our Center is to try to not only comprehensively genotype the people, but also comprehensively phenotype them—and by phenotype, I mean understand functionality at a whole variety of different levels.

We want to understand functionality at the level of single cells; functionality at the level of single neurons; functionality at the level of how the neurons talk to one another—not for "everybody with autism", but for each kid, each individual kid whose genomic sequence you know. We want to see all those steps, how all of these functions are tied together all the way up to EEGs and behavioural tests. I think that we need that kind of annotation in order to be able to understand what's really going on.

HB: And then, presumably, start seeing some commonalities at some level.

JG: Right. What that's going to do is perhaps enable us to separate out different types of autism, just like the way we do breast cancer: some people are breast cancer HER2/neu positive, some will be breast cancer HER2/neu negative. I'm just using these as examples of words, but the key point is that they're markers, they're biomarkers; and that's just something we don't yet have in autism, we don't have any objective biomarkers.

The only kinds of objective tests that we have are these things way out on the behavioural levels; and the problem with that is that we know that trying to make medications at the behavioural level fails.

That's why all the big drug companies have pulled their neuro-psychiatric drug discovery units. They've closed them down, with many scientists let go, because of the failure of a behavioural assay to reproduce something that can cure a human disease.

You're so far away from the genes at that point that the likelihood that you're going to be able to similarly replicate the problem is just vanishingly improbable. That's why we have to be anchored in the genetics. Of course you want to test that the behaviour gets better, but you want to know that you're on the right route towards getting there: there are probably far too many ways to get at these complicated diseases.

HB: The right analogy, it seems to me, is what you were talking about before with migraines: tracing a path down to the genetic level through these ion pumps, or whatever. Until you do that, it's awfully hard to imagine what sort of drug you could develop.

JG: I think, too, that it's important to count on some very powerful mutations. If, for most people, it takes a dozen mutations together, with some environmental stressor to push them to the disease, it becomes highly problematic.

It's much easier to chase the pathway in an individual where having one mutation, with or without the environment, is enough to push them towards that disease, so you can understand what pathway is being perturbed. It's not going to sample everything, but it's at least going to give you a rough outline of how one might make that process happen. It gives you a "cartoon" of pathogenesis, if you will. I think that that's really a very important starting point.

HB: Let's summarize, then: where are we now and how much further do we have to go?

JG: Where we are now—and this was the argument that I made to the William & Nancy Thompson Family Foundation, who generously responded by putting together an entrepreneurial investment of phil-anthropy in our Center—is that, while we certainly don't know all the genetic components that go into the disease, and we certainly don't know all the mechanisms that are involved, we know enough now, genetically, to be able to begin targeting a pathway that does involve the disease process of autism. And it would be wrong to not now start moving on that.

The timing is right now for two reasons: the drug companies aren't doing it—they've been closing down these kinds of units—and we think we have a paradigm that makes a lot of sense, a very compelling and different paradigm to build up from this cellular level.

I don't want to wait until we've found all the genes that are involved. We've got plenty. We can put them together and see a very compelling story for how this calcium signalling is involved.

We know how to intervene in these kinds of things. We know how to make molecules that play a role in how these kinds of signals are involved. We have very sophisticated technology. At this univer-sity, we're blessed with wonderful scientists with huge expertise in looking at these issues. What we need to do is get those individuals who are working on a variety of different problems to all share their expertise aligned to autism, which will give us a very powerful discovery pathway that takes us all the way back to the genes and all the way up to human behaviour.

The Thompson Family Foundation was willing to invest in that. We told them that we felt that we had a shot at aggressively approaching and trying to cure—and that's still what the plan is: that we'll get things into clinical trials.

A paradigm that was really important for me, which might seem at first completely unrelated, has to do with the new discovery that happened in cystic fibrosis. That was led by someone who had gone through our training program, who's done a brilliant job of really moving towards curing cystic fibrosis, but he's done it in a completely different way than we typically approached things.

Just like with autism, it's hard to know what the animal phenotype was: there was no animal model of cystic fibrosis. Now, cystic fibrosis involves just one gene—I don't mean to say that the genetic complexity of cystic fibrosis is the same as autism—but if you change that gene in a mouse you could see some things, but you wouldn't get the clinical disease that affected humans, so you couldn't make a clinical trial based on animal model data.

So they went back and said, *"We're just going to show you that we can cure a specific, broken molecule in humans; we're going to cure that broken molecule. We're not going to give you an animal model, we're going to go right in and fix that human cell; and we're going to take one mutation that we really understand very well."*

There are some 2,000 mutations that lead to cystic fibrosis, but they decided to pick one—a rare one, not the common one that leads to cystic fibrosis—to give them a molecule that they could screen.

They screened thousands and thousands of molecules because they could do it at a cellular level—there are all kinds of automated ways of looking for this. They were able to get a molecule that fixed the effect of that mutation; and then they were able to take that to clinical trials through the FDA.

Now, that's a paradigm shift. For many years people struggled, trying to get an animal model, but suddenly you could go to clinical trials without that, by dealing directly with the relevant lesion in a human.

It also was a paradigm shift in terms of this idea of entrepreneurial philanthropy: the Cystic Fibrosis Foundation sponsored them to try to do this. The surrounding research culture used to be a bit more skewed towards broad-based understanding, but they finally said, "*We really just want to fix this disease, somehow; we're not that interested in general matters of how things work.*"

It got FDA approval very quickly, and it's been zooming along. It clinically targets, not the symptoms of the disease, but the cause of it. And that's where I think we have a window: the genetics in autism gives us a window into the cause. We have some very strong molecular players; and those are the kinds of things that I think we can, and should, be targeting to try to fix the disorder.

Questions for Discussion:

1. Why do you think that there is still considerable resistance to the idea that childhood vaccines do not cause autism?

2. Do you think that autism can be categorized as "one condition" or "many conditions"? What sort of potential medical discovery might change our belief about this question one way or the other?

V. Pathways and Pleiotropy

Searching for mechanisms

HB: When did this cystic fibrosis breakthrough happen?

JG: Only two or three years ago. It was a huge breakthrough; and there was just a very recent triumph, if you will, where they were able to take it and apply it to the common mutation in cystic fibrosis.

The first set of trials only goes back maybe three years, at the most. But it's in humans now: it's an FDA-approved drug now for humans. So, you really can do this.

HB: That's remarkable, not only in terms of what's been accomplished, but—as you say—it's also remarkable in terms of the paradigm shift from the FDA's perspective.

But I would think that, somewhere along the line, the drug companies would recognize that there's a new and different way of doing things, which might cause them to be more enthusiastic about the sort of work, if only by analogy, if nothing else.

Has there been a sense that people are saying, "*Oh, you're trying to do to autism what these guys did to cystic fibrosis, at least roughly, at a schematic level. Sure, there's more complexity, sure there are more genes at play and all the rest, but now there's a precedent in terms of efficacy*"?

JG: Two years ago, Roche invited fifty scientists from around the world to come to a meeting at their castle in Switzerland to discuss whether or not there was enough genetic information to move forward towards addressing autism.

At the end, they concluded that there was; but I was actually surprised because, at that point, this first cystic fibrosis breakthrough had happened and only a handful of the people there were aware of that. Well, perhaps that makes sense, because why would you know about cystic fibrosis if your research is in a neuropsychological field like autism? But still, I was a bit surprised, because for me it was like a bolt of lightning. I'm not sure how many people believe that there's a connection between the two at all but, for me, it's very important.

HB: Structurally there's obviously a connection, based on your whole approach.

JG: Well, that's how it looks to me, naturally enough. But I can't guarantee that my approach is the right one.

HB: Let's talk a little bit about that. What would a sceptic say? Tell me about the prospective criticism from an expert in your field who would say, "*You're misguided, you're deluded, that's not the right way to look at things.*" What would his argument be, exactly?

JG: Well, the argument would be that cystic fibrosis is easy: we know you've got only one gene at play.

And of course that's right, there's only one gene. But they've known about that gene for a long time. You have to know how to approach it, you have to know how to fix it.

HB: So the argument would be something like, "*OK, the methodology works, but it's much more simplistic in this case of CF where you can isolate what's required.*"? Could another argument also be, "*There's no way you're going to isolate the panoply of higher-order numbers of genes in autism,*" or "*We'll wait until you do*"? What else would they say?

JG: Well, again, it's worth emphasizing that I don't think that we do need to do that.

I think there are two important steps. First, there are important correlations with diseases like the mitochondrial diseases. As it happens, I've seen a lot of kids with mitochondrial diseases over the years. That used to be a separate part of my life, but now they've kind of converged together as we've come to understand more what roles mitochondria are playing in the cells.

It turns out that mitochondria play a critical role in calcium signalling; and I believe now that a lot of the lesions that we saw in mitochondrial disease that contributed to autism are, in fact, sharing mechanistically this whole problem of calcium homeostasis. I think that message is becoming clearer and clearer.

Handling these mitochondrial diseases had the same problem: a very, very, difficult diagnostic situation. When we started being able to do skin biopsies on these kids and actually being able to measure things at a cellular level, we were able to sort them into different categories, we were able to understand the diseases.

I believe that getting some functional biomarkers is going to be critical to allow us to make diagnoses: sorting people into groups of those who have autism and this biomarker and those who have autism and don't have this biomarker. That gives you a more coherent group of people who will be going into clinical trials.

It also has a very important implication for the discovery process, because the thing that you're able to use as a biomarker winds up potentially being the thing that you're able to use to screen for drugs that fix that biomarker situation, or to screen for things that make that biomarker situation worse, such as some environmental challenge.

I think that's a really important step. Meanwhile, a lot of people think that the place to look is how the wiring diagram in the brain happens—but, of course, we don't do brain biopsies for obvious reasons. So, that's not going to happen. Moreover, if that's the place where we need to find the difference, it's going to be a really tough job.

HB: Well, that could be an effect, right? I mean, that could be an effect from all of this.

JG: Downstream, exactly. That's my sense. I'm sure that all that is found there, but I think it's too far away.

Most of the arguments that have been made in model systems are at the level of doing brain slices and really looking in-depth at the brain and how it behaves. But I believe that there's a real value in looking at the simpler, core level before you get to the complexity of the brain. My sense is that if we're going to be able to have any success in doing this at all, that's the way it's going to happen.

HB: That's your "physics side" talking—your biophysics side.

JG: Yes. We don't have to find all of the genes, we have to find an actionable gene, an actionable pathway; and that's what I think the calcium story is pointing us towards. It's also important in that it's quite similar to the mechanisms that are involved in cystic fibrosis. It sounds like they're a million miles apart: one's in the epithelia and one's in the nerves, but it's also important to recognize that autism isn't only a neurological condition. The kids have a multi-system disease.

That's something geneticists are used to. We're used to the fact that gene defects cause pleiotropy—they cause a lot of different problems in different tissues. We run into that all the time. And I would say that virtually every kid who has autism also has some problems with their GI tract, lots of them have problems with their immune functions, and so forth.

So if you think that you're going to be able to solve the problem of autism just by looking at synapses in a specific region of the brain...

HB: What about all this other stuff?

JG: Exactly: what about all this other stuff? Once again, then, it's naturally appealing for me to be looking at a very fundamental mechanism as being involved.

Now, to your other question: "*If it's involved in everything, how can we move forwards?*" Well, going back to these association studies, what we find is that we're not so good at "naming diseases" in the first place. It turns out that the genes found in bipolar disease (what used to be called manic-depressive disease, now bipolar disease type 1) are very similar to the genes that are involved in schizophrenia.

Ten years ago, psychiatric geneticists would adamantly maintain that those diseases have nothing to do with one another, but as we learned more and more about the genetics, we saw more and more similarities. What our eye calls "bipolar disorder" or "schizophrenia" probably turns out not to be a fundamental distinction. And autism is in that same cluster.

From my perspective, it's not that I don't know the names of the different neuropsychiatric diseases, it's just that I think that what you wind up with is a hyper-excitable neuron—it's that kind of a problem.

And the reason why I use that kind of hyper-excitability analogy is that we really understand these kinds of diseases in the heart. In the heart, it causes all kinds of arrhythmias.

The paradigm that I've been using in review articles trying to explain my approach involves looking at something called the "Long QT syndrome". In the first review that I wrote, I said that it was curious that we don't see this one kind of calcium channel ever mutated: it must be lethal.

Well, suddenly, a mutation in that calcium channel of the heart was found to cause this arrhythmia—it's called "Long QT"—and 80% of the kids who have that mutation have autism.

There are 12 genes, which are all involved in calcium homeostasis, which all cause a specific form of arrhythmia called LQT. It's a lethal arrhythmia, so it stands out. They're all dominant, so you can follow them in families through generations.

And sure enough, there's this one rare mutation called Timothy's syndrome—there have only been a handful of cases—and this mutation is associated with not only this arrhythmia (which we understand mechanistically, because we know all the players), but more significantly you also see a highly-penetrant lesion in how the neurons are working that give you what looks like autism.

To me, that's one of the most powerful messages in the autism story. This doesn't necessarily tell us all the ways you can make autism, but it at least tells you one way to do so.

HB: Wow, that's a very strong correlation.

JG: Yes, it certainly is a "wow" kind of thing. I could talk a lot more about that lesion, but it's a real touchstone for people who look at biophysical signalling mechanisms to have such a highly-penetrant mutation—for a geneticist, "highly penetrant" means that, if you catch that mutation, you're going to show the manifestations of that mutation. And to have a mechanism that we understand very well, like this arrhythmia, which is quite a specific kind of lesion (you can only break this channel in a couple of ways to get this to happen, it's really remarkable) yields a very powerful model for how we should move forwards.

HB: And again, methodologically, it also shows why it's important to look very broadly. You never would have imagined that you would have some correlation for autism and manifestations of a heart arrhythmia.

JG: That's right: you certainly have to have a broad perspective for autism; and I think that's very powerful about what our Center is trying to do. We have 60 different investigators involved who really span the gamut and have expertise in a variety of different places. When you ask them to just use what they're doing and think about autism with that, it becomes a very compelling discovery platform.

HB: Are there other Centers that are doing something similar? How unique are you guys at CART?

JG: I think that we're very unique. People don't necessarily tell you everything that they're doing but, by and large, typical autism centers are based sort of on the NIH idea that you're just going to do a lot of good science.

For us, having the Thompson Foundation give us a strong mandate is a way of pulling people together: we're able to seed cooperation. Faculty don't usually do things like that, and industry couldn't build up that strong of a platform on its own: to be able to hire enough people to build that kind of a strong platform would take a very long time.

So, I think what we're doing is unique. Certainly when we were starting to do it, it was unique. Maybe other people have started doing similar things by now, but I don't really know.

HB: There is a view amongst some members of the community that looking at autism as a "disease", or a condition "to be cured" is inappropriate and demeaning. According to this view, one should just look at it as just part of the variation of the human experience. This is not my view, as it happens—nor is it yours, I understand—but I'm sure that there is a need to be sensitive to those who do have that view.

JG: Yes. Tom Insel, the director of the NIMH, addressed this issue during an autism summit we had in California recently, and it is certainly something that I have encountered. Inevitably, whenever I give a public lecture, there will be a couple of letters in the newspaper saying something like, "*How dare Dr. Gargus propose to cure autism! Why make me like everybody else? I'm fine the way I am.*"

I think there are a variety of issues around that. We know that the social cost of autism in the United States is well in excess of 130 billion dollars per year, to say nothing of the emotional turmoil: it's a major societal issue.

I think the idea of trying to be able to come up with medications that could ameliorate the disease makes sense. I don't think anybody is going to be mandating that people be treated, but what Dr. Insel said is that in some cases prospective legislation to get funding for autism projects was killed off by this contingent, who were basically saying, "*Don't do that.*" It's a civil rights issue."

HB: Well, I'm not an expert in civil rights, nor do I pretend to be, but I could imagine that if we did find ourselves in a situation where there was a cure for autism—or there was some medication that would vastly ameliorate the symptoms, or whatever language you want to put around that—the civil rights aspect presumably wouldn't impinge on whether or not people would have to use it, which is a whole different issue. If I have a headache, I can go to the pharmacy and pick up some Advil. It's there; and I can elect to take it or not.

JG: Well, yes. Again, I really can't fully appreciate the whole spectrum of this. I know that it's something that, as geneticists, we've encountered in Little People of America, in something that used to be called different kinds of "short-limbed dwarfism syndromes". It's also something that we've encountered in the deaf community in a number of the hereditary hearing disorders.

There are cases where, for example, a couple that has achondroplasia—the common form of short-limbed dwarfism—will elect to terminate a normal pregnancy because that's not what they would like; they would like to have another child who has achondroplasia.

You can wind up in situations that are way above my pay scale—I just can't figure out how to sort all these kinds of things out—but I feel that autism is something that warrants our trying to intervene; and I think to use the phrase "trying to cure" autism is not inappropriate. I certainly try to think about the sensitivities of the community, but I can't see that this justifies us stopping trying to get to the bottom of this.

Questions for Discussion:

1. How would you respond to the claim, "The whole concept of 'pleiotropy' just means that we don't know enough yet. Once we have a complete under- standing of the underlying mechanisms at play, pleiotropy will disappear"?

2. To what extent can we objectively distinguish between "highly penetrant" and "not-so penetrant" lesions? Might the level of penetration depend on other factors?

3. Should autism be regarded as a "disease"?

VI. Reasons for Optimism

Streamlined bureaucracy and bold prognostications

HB: So, continuing with the theme of public policy that moves beyond your current responsibilities and expertise—since I like to push people out of their comfort zone—if you were President of the United States, what sorts of things might you do, or might you do better, to lay the groundwork for a faster road towards reaching some of these conclusions that you hope to reach?

JG: Wow. No one's ever asked me that before. Well, the FDA has become quite responsive, specifically in the context of what are called "rare diseases". There are mechanisms to get clinical trials—and, again, this is something that's guided us in terms of trying to decide what's feasible—because now, for rare diseases (which a lot of the subtypes of autism that I've described to you fit under) you can do a small "proof of concept" clinical trial which is like an investigator-initiated clinical trial; that's something we intend to use.

Now, it's not good enough to get you FDA approval, but what it can do, with a relatively modest sample size that could be affordable in an academic setting (it can only be done at an academic institute, drug companies can't use this mechanism), is that if you get something that's promising then you can find a partner in the drug company. The drug companies aren't going to partner with you until you can show them that you have some kinds of efficacies.

The FDA is trying to be responsive to the situation. There is still too much of an administrative burden: just being able to get all of the Institutional Review Board pieces in order to carry out our clinical trials typically takes a year or so; it takes a long time to go through all those hoops.

It would be nice to expedite that. It would be nice, too, to find a simpler mechanism to make what's an approval process at one institution work in another institution, so that you don't have to reinvent the wheel every time.

A lot of the bureaucratic impediments and concerns are getting recognized—the NIH and the FDA, working together, recognize that these are problems. Nobody wants to slow this down: Dr. Collins and I were fellows together at Yale at the same time, and I know he's a smart guy who wants to fix diseases.

Nobody's trying to make it harder: they're trying to make it easier. You have to do it in an ethically-sensitive way. Another initiative that is being pushed, takes us back to what I was saying earlier about how we've never been naming diseases the right way. We've been using these make-believe things thinking that we know what the mechanism is, but we should be working towards naming the diseases on a gene basis. That's what we did in bacteria a long time ago: we didn't bother naming the bacteria after how they looked, we named them after the gene that they had.

If we begin talking about these diseases with a more explicit mechanistic framework, that might simplify a lot of things. A whole host of diseases have these complicated architectures. If we start calling them the way we call cancers—a HER2/neu positive, say—if we start using those kinds of gene-based markers to define what the disease is, we may come closer to having coherent studies.

And obviously the more coherent your study is, the more likely that a clinical trial will succeed. Part of the big problems in the clinical trials are that they get all the way up to a human trial and then the whole thing falls apart: it just doesn't work. And it costs hundreds of millions of dollars to have a trial that fails, so we can't do a lot of these.

What we really need to avoid is a "dirty sample": you don't want to be mixing apples, oranges, bananas and grapefruits. You want to be comparing bananas to bananas, you want to make sure that you're looking at a similar disease when you're doing a clinical trial in order to see if your medication is helping those people.

I think that all of these improvement are going to help. Obviously, everybody would agree that more money and more continued support would help a lot. The NIH budget has been unbelievably savaged through the years: the paylines are discouraging to trainees, they're discouraging to the people who are in the field and they're making people do things in completely different ways.

When I was on study sections, you could expect that the payline would be 25%, which means that we'd fund a quarter of the grants that came in. Now the paylines are down to around 5–6%; it's almost at the point where an investigator will ask herself, "*Am I really going to bother doing this, or should I just go out and buy a lottery ticket instead?*"

What that's also done is naturally drive people to ultra-conservative approaches. If you know that only a tiny sliver of the grants are going to get through, you're not going to take a flier on somebody who has a very different idea; instead you're going to take the most conservative, putting one brick on top of another, kind of approach. And that probably isn't going to take you anywhere.

I think that's changed the psychology behind how the grants process works. Where is that money going to come from? I have no idea; and I also recognize that that's always the answer that people give. It's a very self-serving kind of answer for a scientist to say, "*Oh, you should give more money.*"

HB: What, very specifically, would you do with more money? If you had more money, if I could triple your budget here, would you be hiring more people? Would you be conducting more extensive experiments? Is there a clear sense of, "*Gosh, if I only had more funds, I could do this specific experiment*"?

JG: Yes—all of that. And also, as I said, I think the psychology of the funding made it harder for us to have a group of graduate students, a group of postdocs that are working on a project.

Now, we've tried to do it by putting together a Center where we have different groups working together. We try to simulate that kind of thing, but I think at each individual investigator's level, we shouldn't be closing the doors to these folks.

If you don't have enough money to pay them today, they're not going to be that interested in pursuing that approach because they'll naturally ask themselves how they will be in a position to keep doing it in five years, and you're going to have a problem. As I said, I think it changes the psychology of the whole field.

HB: OK, time for some major speculation now. I'm not going to hold you accountable to this, but I'd like to know your view on how many years from now we'll have to wait before we're going to have some kind of a breakthrough in autism that's analogous to the cystic fibrosis breakthrough.

JG: For sure in five years but I'm hopeful in three years. By that, I mean at the level of a clinical trial. I don't mean a bottle on the market shelf.

HB: That's completely fine. A clinical trial is still a pretty big breakthrough.

JG: Right. I think a successful, clinical trial in 3–5 years is not at all unreasonable, and that's the ballpark of what we had promised. I don't think it's crazy at all—but that's just me, of course.

HB: Well, I'm talking to you. I can't expect you to answer for anybody else.

JG: Of course, if you sampled everybody who's working in autism, many would say something different; but that's what I believe. I don't think that that's an inappropriate window.

HB: Great. Anything we've missed?

JG: No, we've really covered a lot. I don't have any other points that I wanted to bring up.

HB: Well, thank you very much, Jay. This was great.

JG: Thank you.

Questions for Discussion:

1. How might the funding allocation process be adjusted to encourage people to develop more innovative and fundamentally different approaches?

2. Will we ever have a comprehensive and effective treatment for autism? If so, when? If not, why not?

Learning and Memory

A conversation with Alcino Silva

Introduction

Dom Alcino and the Age of Discoveries

When I learned, shortly before speaking with him, that Alcino Silva had received the 2008 Order of Prince Henry award for his contributions to neuroscience, I must admit that I didn't think much of it.

After all, governments give awards all the time; and the fact that Silva, who grew up in Angola before moving to the United States, had his work recognized by the Portuguese government, hardly struck me as anything particularly noteworthy.

The truth is that I had never heard of the Order of Prince Henry. In fact, I had never heard of Prince Henry.

But it turns out that I was missing something. Because Infante Dom Henrique de Avis, the 15th-century Portuguese king commonly known as "Henry the Navigator", was one of those rare visionary monarchs who was keenly aware of the possibilities of his era: under his astute administrative leadership, the so-called Age of Discoveries, and the Portuguese Empire, began.

For Alcino, however, the most impressive thing about the Infante Henrique wasn't so much the founding of an empire, but his farsighted recognition of the powers of human ingenuity and how it could best be harnessed.

"This is a man who established what we now know as an institute at a time when everything was really balkanized. He paid a lot of people to come together and develop navigation, which was very unusual. He gave them large amounts of money and land, and all they had to

promise was that they would share everything they knew with each other, which was unheard of.

"Then we went elsewhere—and some of that history is checkered, as you know—but it changed the history of Portugal, and it started this concept that you could develop instruments and share knowledge for the greater good."

Alcino runs a learning and memory lab at UCLA that is focused on a vast number of topics, from schizophrenia and autism to memory enhancements and aging. One of the founders of the field of molecular and cellular cognition, he and his colleagues focus on understanding the specific molecular mechanisms of neurobiology in the hopes of being able to intervene and repair these mechanisms when they go awry. And even attempting to do such a thing naturally requires a more generalized approach.

"There was a time in neuroscience, twenty or thirty years ago, when knowledge was effectively sealed off into separate compartments: there were molecular neuroscientists, psychologists, and physiologists; and these groups hardly talked to each other.

"Then there were a number of technologies that came into play that allowed these groups to interact and to have something to talk about, to be able to do experiments together. That brought them together. And that is actually the origin of molecular and cellular cognition.

"When I helped form that society, twelve or thirteen years ago, we didn't have a community of people who worked with molecules and cells and behaviour. But we really needed that kind of community, because it was different than just working in any one of these areas. There were special questions, special needs, special approaches that we had. Our papers looked different. And that's why we formed that society.

"Nowadays, a great part of the work in neuroscience is work that connects different areas, from molecules all the way to behaviour. But this is a relatively recent change in neuroscience. For most of its history, neuroscience was really separated into different fields."

Suddenly it is no longer about some award—it's hard to stop oneself from contrasting this dynamic UCLA neuroscientist with Henry the Navigator himself. Indeed, the fact that he is keenly engaged in the pursuit of charting the current scientific landscape so as to develop, as he calls it, a "Google Map" for neuroscience, only makes the comparisons between Alcino and his 15th-century compatriot all the more striking. Perhaps, one might think, there is a Portuguese gene for cartography.

But what, specifically, have we learned from all of this restructuring? What's been discovered?

> *"We have recently amassed enough evidence to now appreciate that, during learning, we change the synaptic weights—how neurons communicate with each other. These changes in synaptic weights are orchestrated by hundreds of molecules, and these molecules regulate these changes in cell-cell communication in the brain, which, in turn, regulates learning and memory. Molecules that trigger these changes are involved in learning, molecules that maintain these changes are involved in memory."*

The experiments speak for themselves. In Alcino's lab alone, which works with mice, specific memories have not simply been localized, they have been precisely manipulated.

> *"In one of our experiments, we have given the animals two memories: first, that a salty substance was not so good because it made them slightly sick, and the second, that there was a tone that was to be avoided because when they heard that tone, they got a buzz, a slight shock.*

> *"What we did then was to change the physiology of the brain in such a way that we determined where one of these two memories went to in the brain, but not the other. We let the other just go into the brain normally.*

> *"And what we were able to do was to get rid of one memory, but not the other. We can manipulate memories in animals by selectively inactivating one memory and not the other. Then we let the animal*

recover and that memory comes back again. We can literally turn the switch on and off on memories now with the special tools that we have designed."

If you're finding this all too much like science fiction, and not a little unnerving, it's time to unveil some unequivocally positive news from the neuroscientific front lines.

Alcino and his colleagues have explicitly and repeatedly demonstrated that it is possible to reverse the cognitive deficits, the learning and memory deficits, of an animal model of something called neurofibromatosis type one, or NF1 for short, a genetic disorder that is responsible, in 30-40% of patients, for a wide range of learning difficulties involving memory, spatial navigation, attention and motor coordination.

That Alcino and his colleagues have been able to coherently construct and analyze mouse models of this condition is impressive enough. That they've found some pharmacological treatments that allow the animals to cope better is inspiring. But actually **reversing** cognitive deficits through drug treatments? That was truly revolutionary.

Because until recently, the common understanding was that cognitive deficits resulting from developmental disorders were simply **irreversible**: that once a brain had been damaged in the development process towards adulthood, there was no possibility of making any fundamental repairs.

"These results meant that adult animals that were unable to learn very well, that had problems with spatial navigation and interaction with their environments compared to other animals, could have their learning deficits reversed when we addressed the biochemistry that was affected with a specific drug.

"But the results also demonstrated something that we should have known already: that there's a tremendous amount of flexibility and plasticity in the human brain, and we systematically underestimate the brain's resourcefulness at recovering and repairing itself.

"This has been one of the really important things that we found—something with the greatest potential for human impact—it opened the door to the possibility of treating literally hundreds of millions of people worldwide with problems that we never thought we would ever have even a chance of touching.

"I'm convinced that our children, certainly the children of our children, will have a very different relationship with these types of disorders than we have with them today. Think about the time before antibiotics, how the world was: a simple cut could kill you. Now, we don't even think about it, we cut ourselves, sometimes severely and we just wash it, treat it, take antibiotics, and most of us easily move on.

"Just imagine a time when for parents of a child with neurofibromatosis, with autism, with schizophrenia, with any of these horrible, horrible disorders, it will be like a cut—something you need to address, something you need to treat, but hardly a life-changing condition like it is now."

Just imagine. Truly, a sparkling new land to steer towards.

The Conversation

I. Planting Seeds

Laying the groundwork for future discoveries

HB: There's an awful lot I'd like to ask you, but I'd first like to talk a little bit about your origins, how you got into the field. My understanding is that you're of Portuguese descent, but that you spent much of your formative years in Angola.

AS: Yes. My father went there when he was young—the most adventurous of the Portuguese went to the colonies. It was a wonderful place. I left when I was eleven because of the war. Then I spent another few years in Portugal before coming to the States. But Africa was amazing. I have not been able to go back yet, but my father tells me that it's safe now, so I'm planning to go back. I have "google-visited" the place quite often, spending a few hours navigating around, trying to find places that I recognize.

HB: Is your old house still there?

AS: I haven't yet been able to find my house, actually, which is strange. I don't have a very clear memory of where it was relative to big landmarks. I know the general area, but unfortunately I haven't been able to locate it yet. Luanda is a beautiful city by the ocean. I have great memories of it, but when the war came, everything changed.

HB: So, you came back to Portugal at the age of 11 to start high school, after which you left for the United States, to Rutgers, for your undergraduate degree. Was it a difficult transition when you first arrived?

AS: No. I was so excited. I missed my parents, I missed my girlfriend. But I was so excited to be here. I wanted to leave Portugal at the time because the universities were so disturbed by the revolution in 1974 that disorganized the country. There was a great deal of upheaval that affected universities, and I used that as my excuse to essentially convince my parents to let me go away to the furthest place I could go. I had hitchhiked all over Europe, but it's hard to hitchhike to the States.

HB: They needed some convincing?

AS: Well, you know, all parents need convincing to part with their child. They have to see with clarity that the place he is going to is good for his future. It was difficult for my parents: I didn't go home every few months. They knew what it meant for me to come here: that, eventually, I would probably end up getting married and staying here, which is exactly what happened. So they knew that, and it was difficult. But I have done my best: I go back home frequently, and I've had associations with Portuguese universities, so I do the best I can to pay back the country and visit my family.

HB: And in 2008, I understand that you were given a very prestigious award from the Government of Portugal, the Order of Prince Henry.

AS: Yes. I was very pleased, especially because the award commemorates one of the most inspiring figures in history that I know, Henry the Navigator, the principal initiator of the Age of Discoveries.

This is a man who established what we now know as an institute at a time when everything was really balkanized. He paid a lot of people to come together and develop navigation, which was very unusual. He gave them large amounts of money and land, and all they had to promise was that they would share everything they knew with each other, which was unheard of.

HB: The original open source.

AS: Yes, that's right. And they did. They came and formed this institute and he changed the history of Portugal. Then we went elsewhere—and some of that history is checkered, as you know—but it changed the history of Portugal, and it started this concept that you could share knowledge for the greater good.

In fact, it was not simply sharing knowledge: there were also instruments that were developed as a result of these collaborations, because of the Infante Henrique. And the award that I received was commemorating this man whom I admire deeply.

HB: It was all very fitting, then. I hadn't appreciated that there was a tremendous amount of symbolism of this award in terms of utilizing science and technology, spreading our understanding for the greater good. I had assumed that it was just another award.

AS: Yes. I think it's one of Portugal's proudest moments, actually— that, together with another incredible event, where this other king, Dom Denis, planted a forest from the North to the South of Portugal in the late 13th and early 14th centuries. Can you imagine the enormous amount of foresight for that, with the vision of developing navigation later on? Imagine the vast resources that you'd have to invest a generation ahead of actually getting any returns. That was Dom Denis.

HB: He was already thinking about navigation back then?

AS: Yes. Because for navigation you need boats, and for boats you need wood, and for that you obviously need forests. We didn't have enough wood in Portugal.

Another problem was that we had erosion from the wind and sea spoiling some of our land. But that wasn't really the big thing. The big thing was to plant trees and grow wood for the next generation. Just imagine if our politicians could do that today: plant wood for the next generations.

HB: It's often hard to get them to see a month ahead, let alone an entire generation.

AS: Exactly. And this was an enormous amount of resources—if I'm correct, half a mile wide through a long stretch of the coast of Portugal. Those woods are still there today, seven hundred years later.

And then later, when we had all this wood and all these boats, came Infante Henrique to develop the science that allowed us to circumnavigate the globe, to go around Africa and South America. I mean, we just went everywhere, you know, because these two men had the vision to do so. It's really incredible.

We often think about history as disconnected from us, but it has an inspirational side to it: men who are related to you were able to do these types of things. I wonder how many young men in Portugal have thought about that. I think it's important.

HB: Indeed. And when you came to the United States for your undergraduate degree, my understanding is that you were interested in science, but you were also interested in philosophy—you were interested in epistemology, where knowledge comes from, and so forth. Eventually, you were able to reconnect these things, but I imagine that they were naturally quite separate at the very beginning of your university education.

Did you have these ideas when you first started? When you left Portugal were you thinking "*Oh, I really love science, I want to do science, I want to do something in the scientific world,*" or was it just a sense of moving away during a tumultuous time and having some new and exciting experiences?

AS: Well, young people love adventures, so that was a big component of it. But another one was science.

HB: So you were already strongly motivated to do science from an early age?

AS: Yes, I was.

HB: Did you have a sense of what sort of science you wanted to do back then?

AS: Not really. But science really motivated me; and the question of knowing how we know was always at the center. When you're young, you don't think of things the way you eventually do, but in Portugal people took philosophy classes in high school, which is really interesting. I had two years of philosophy in high school; and one of the questions that fascinated me the most was, *How do we know what we know?*

It was clear to me that this question was central to all of science, because, if we don't understand *how* we process information about the world, how can we really know that what we *do* process is reliable?

Maybe there's an evil deceiver at every step that shifts what we find and changes it in a way that's actually not reflective of reality, or maybe there are systematic errors of processing that we commit that change how we interpret the world and how we engineer things around us.

For me, I thought that if there's one question in science, that's it. But, you know, people know things with different degrees of clarity. And I don't think I knew that with the same degree of clarity that I now express today.

HB: Well, I hope not. I mean, my goodness, if you already knew all that when you were an undergraduate...

AS: I think that was always in the back of my mind. When I was in school, the classes I took were science classes, biology classes— because, again, it's about the brain, it's about the body: *How does this work as an engine of knowledge, essentially?*

Then it became about philosophy and epistemology, and that continued in my graduate studies. Eventually, I decided not to go into philosophy, simply because I just loved science. I loved tinkering in the lab—going to the lab and making things happen was great fun. But then in graduate school, my interest in both continued, and it continues to this day, actually: I was at a conference recently about how we can devise tools and strategies to deal with the immensity of information in science, so all of this goes back to my high school years and to my philosophy classes.

It's funny how we are one person who just morphs slowly, but never really actually changes.

HB: Well, it gets deeper, maybe.

AS: Maybe, yes.

HB: There are many obvious links here to neuroscience: *How do we learn? How do we remember things? How can we be sure that we know what we know?* These things are, of course, all deeply interconnected.

Later, I hope to talk a bit about your work on research maps and the need for developing a clearer sense of what the research world is doing so that we can best capitalize on knowledge and apply it: how we can be certain of it, how to avoid being misled.

These seem to be fairly constant themes that run through your research career and your entire scientific and philosophical outlook.

AS: Yes, that has been a passion of mine. Lately we have focused a great deal on trying to leverage this knowledge to help people who are not born with brains like ours, or something happens and they develop problems with cognitive function.

But it's all part of the human experience—*How do we find our place in this vastness?* For me, my place has been about focusing on, *How do we know what we know?* and, *How can we leverage this information about the brain to help, even in modest ways, those around us?* That has been the focus.

Questions for Discussion:

1. *Do you find Alcino's blend of scientific and philosophical interests unusual? How much do you think this is a product of his exposure to philosophy in high school?*

2. *Should **all** science students receive at least **some** exposure to philosophy during their education? Readers interested in this notion are referred to the Ideas Roadshow conversation with Princeton University physicist Paul Steinhardt, **Inflated Expectations: A Cosmological Tale**, where he makes precisely this argument.*

II. E Pluribus Unum

Exploiting cross-species similarities

HB: I'm naturally anxious to talk about the specifics of your research—about memory, learning and cognitive deficits, and how we can, remarkably it seems, consider concrete ways to be able to improve matters.

But first I'd like to address a more general issue. You've worked a lot with mice, and you've also worked with flies. But for me, there's always the question of how, exactly, this work on other animals relates to humans.

Why is it that mice seem to be the animals of choice for so many of these types of experiments in neuroscience? Is it just because it's easy to somehow manipulate them, or is there a really deep, structural similarity between mice brains and human brains?

AS: It's both actually. It always amazes me that, even when we look at higher-order behaviours and their associated mutations, there might be a link between humans and other creatures like mice.

Take autism. If you have autism, you have problems with social interaction, you have problems with repetitive behaviour. These seem to me to be very human-like qualities, and it's amazing to me that the same mutation in mice can recreate some of these same behaviours.

Here you have a human brain and a little mouse brain, but the engineering principles that were used by evolution to build both are similar. As we look at the genes, as we look at the cells, as we look at the structure of these brains, we find more things in common than not, actually.

You referred earlier to a little bit of work that I did in flies when I was an undergraduate. An amazing thing is that, when you take these brain genes that are known to be involved in memory—genes that are active in memory and are needed for memory—and you look at them in flies, you look at them in C. elegans—these little tiny worms—you find that the same genes, the same fundamental chemical processes are there in these worms when they learn, in flies when they learn, even in Aplysia—sea slugs—they're there too. And you see them in mice, and you see them in humans.

It's really magical, actually. It's really, truly magical, because we sit in this privileged position of being capable of Shakespeare and Beethoven, but so much of what's underlying this you can find, actually, all the way down to yeast.

Yeast cells have some of the same proteins that react to changes in their environment. They communicate these signals, and then the cells do something metabolic that's appropriate. And you see that in flies, in Aplysia, all the way up into humans.

At the genetic level, the structural level, the psychological level, we have conservation. And that's why we can use mouse models—with some caveats, of course—in some studies in humans.

Ethically, it would be just impossible to do the types of studies we need to do, to understand the precise mechanism of how a gene might cause autism, or schizophrenia or Alzheimer's.

We need to know, in a very detailed manner, because we need to intervene and treat these horrible disorders. So one way to do this is to use mice. It's not the only way, but it's certainly one way that has gotten a lot of play, and has a lot of grounding.

HB: What are some of the other ways?

AS: Well, people have used more complex animals, all the way up to primates. They've used other model systems. There are certain aspects of the chemistry of the brain that you can capture even in cells. Neurons change when they react with other neurons, and some of these changes—some of these molecular engines, you can capture just in cell lines. The problem is that in those cell lines you can't capture the other complexity that's important.

HB: Because you've just isolated one mechanism, I'm guessing.

AS: That's right. But all of these models play an important role. Each brings a uniqueness, an advantage, that the others don't. The great thing about cell lines is that you can have a million of them, literally, in little wells, and you can screen through them, looking for things that might give us a hint that a certain drug might work for autism or Alzheimer's or schizophrenia, whereas screening a million mice would be very expensive and very difficult.

So each model brings a different thing to the table, and the goal at the end is to understand and be able to treat these conditions.

HB: Just a brief question about actual experimental procedures. When you're talking about a fly—conducting experiments with a fly's brain—what are you actually doing? I mean, these are really very small things...

AS: They are, yes. In both mice and flies, and other organisms now (rats for example), it is possible to tinker with their genome. The genome is sort of the blueprint of what that organism will become, the instructions that are used by the organism to build a brain, to build a repertoire of behaviours—including memory—that the animal will have.

So by tinkering with the blueprints—by tinkering with the software, if you will—we can then shape the hardware, we can shape the brain.

HB: How do you actually do this?

AS: There are a variety of ways that we can tinker with the genome, and we can do it at different stages of life of the organism.

For example, in the mouse, you can do it early on when the mouse is just a single gamete, or you can do it when the mouse is just a single cell embryo, or you can do it later on when it's a live mouse by taking a viral vector with specific genetic instructions and placing it on a brain region that you want to understand.

Intervening genetically at different stages gives different consequences. If you have a disorder that you're concerned about that the patient has had since birth, you may want to create this mutation early on because that's what happens in humans.

But if you want to understand in detail a process that's underlying attention or emotion, then you may want to manipulate genes or manipulate other things locally in brain regions that we already know are involved in that behaviour.

HB: How, exactly, do you do this local manipulation? How do you get at it?

AS: Well, we have a map of the brain, essentially. We know what part is involved in working memory, what part is involved in repetitive behaviour, what part is involved in spatial navigation, and so forth. So we have a three-dimensional atlas, and we can place a little pipette, like a straw, right into that area and leave a virus there.

Actually, these days we can do even more, we can activate these neurons by changing the channel compositions of these regions, so that when you shine a light, these neurons come alive or are repressed in their activity.

HB: This just seems like science fiction to me.

AS: It's truly amazing. These days, we can even change the brain so we know ahead of time what cells will have a given memory. We can channel memories into certain brain regions, into certain brain cells, and then manipulate those cells and then see what happens to memory.

HB: Well, I'm going to get there—trust me, I'm determined to talk at some length about memory—but I'm just trying to imagine for the moment what the specific techniques are that these guys are using.

They're in a lab. They've got mice, flies, and so forth. And I think to myself, *"Okay, mice I can sort of understand, flies are already starting to flummox me."*

AS: Well, in flies you can change genes like you can in mice. In flies, you have these elements that are transposons: they move around in the genome. And people have been able to take advantage of them to change the genetic makeup of flies.

DNA is a molecule, and you can place it in certain places and then that molecule changes the instructions that the fly has to engineer a whole, big fly. And those flies mate with other flies, and that mutation is passed along, so you have whole strains that are exactly the same except for one change that you have made genetically.

But these days, amazingly enough, you can place these flies on a machine; and you can image their brains as the fly thinks it is flying around.

HB: No! You can get real-time imaging of the *brains of flies*?

AS: Real-time imaging of the brains of flies as they **think** they're flying. You immobilize the fly, and you change the environment of the fly so that the fly *thinks* it's flying—you collect information about the torque from its wings—and then you can image the brain.

HB: This is really amazing. You're not making this up?

AS: This is really true. See, this is why I have a hard time getting away from the lab—this stuff is magical!

Our group doesn't do flies, but we do things in mice that are just as amazing. But this hasn't always been the case. I remember, as a student, that one of the big problems in neuroscience was how few things you could do.

But, now, everything has changed. Now we have far more technology than we know what to do with. The limitation now is not whether we can do X or Y, the limitation now is whether we have enough imagination and creativity and a working memory and a knowledge base that will allow us to be able to take advantage of this technology and do things that are sufficiently creative and innovative.

We are no longer limited by what we can do, now we are limited by the sizes of our imaginations, which is an amazing place to be, because essentially what this means is that the problem space in neuroscience has been blown wide open. Where you go on that problem space is only dependent on you—which can sometimes be a tough thing for scientists.

It's a very nice problem to have, of course, but it does have its problems: you can't fall back upon saying, *"If only I could do this,"* because now you ***can*** do it. Now it's a case of whether your *mind* can take you there, which is a different question.

It's a really incredible time in neuroscience, there's no doubt about it.

Questions for Discussion:

1. What do you think Howard means, exactly, when he says "Because you've just isolated one mechanism"?

2. Do you think that there will come a day when computer modelling will replace all experiments on animals?

3. To what extent might advances in experimental biology lead to a deeper understanding of what differentiates one species from another?

III. Putting the Pieces Together

The sociology of neuroscience and running a lab

HB: I have some related questions about labels and categories. So, I begin to read about this Alcino Silva fellow, who, I discover, is a "molecular and cellular cognition guy"—MCC. Then I read some more and discover that he is reputed to be one of the founders of MCC.

But I think to myself, *"Well, isn't this just 'neuroscience'?"*—the sorts of things that you've just been talking about: we understand the various mechanisms better at the cellular and molecular level, and we're naturally working with DNA to manipulate it so as to observe patterns of genetic inheritance.

I mean, isn't that the obvious sort of thing that we should be doing? Put another way, what do "non MCC" people do in this field?"

AS: Well, neuroscience is huge now—there's so much going on. There are people, for example, who are interested in understanding the human brain at the level that imaging allows you to understand it.

You place patients in fMRIs, you place them in a PET scanner, and you look at potentials in the brains of patients. And then, from all of this information, you develop models of how brain regions interact during important, interesting, behavioural things that we do: memory, attention, emotion, consciousness and all of that.

This is all part of cognitive neuroscience, which I doubtless grossly oversimplified. It's a huge and very interesting field outside of genes and proteins and cells and physiology and all that.

There are individuals who are strictly interested in clinical neuroscience: understanding patients, understanding disorders—how to diagnose them, how to find those unique aspects worth studying. That's a huge world too.

There are those who specialize in development. There are neuroscientists who focus on understanding how the brain develops, because there are so many developmental processes that go awry and give origins to problems that we see in adults.

The idea there is that if we were to understand in detail how the different brain regions develop, how neurons migrate, how all those mental processes take place, we may be able to have a deeper understanding of the origins of schizophrenia, autism and all those terrible disorders.

HB: Sure, but that's clearly directly related to what you're doing.

AS: Yes, but this landscape is very, very large. There are so many areas that need very careful study.

There's a very successful colleague of mine here at UCLA, Dan Geschwind, who principally focuses on trying to understand how genes interact with other genes, what genes are responsible for autism and other disorders, and so forth. There's a huge world there, too.

What makes molecular and cellular cognition unique is the fact that we work between these fields, if you will.

There was a time in neuroscience—I'm talking twenty or thirty years ago—when knowledge was effectively sealed off into separate compartments: there were molecular neuroscientists, psychologists, and physiologists; and these groups hardly talked to each other.

Then there were a number of technologies that came into play that allowed these groups to interact and to actually have something to talk about, to be able to do experiments together. That brought them together. And that was the origin of molecular and cellular cognition.

When I helped form that society many years ago, we didn't have a community of people who worked with molecules and cells and behaviour. But we really needed that kind of community, because it was different than just working in any one of these other areas.

There were special questions, special needs, special approaches that we had. Our papers looked different. And that's why we brought together that society.

Nowadays, a great part of the work in neuroscience, as you've said well, is work that connects different areas, from molecules all the way to behaviour. But this is a relatively recent change in neuroscience. For most of its history, neuroscience was really separated into different fields.

HB: Is that manifesting itself in the educational process? When one examines what is required for the relevant undergraduate and graduate degrees, how quickly are these changes in the field being incorporated?

AS: They are catching up. I see it in my students. I started from the molecular and cellular end of neuroscience. When I talked thirty years ago with molecular and cellular neuroscientists, they didn't even know the basics of behavioural neuroscience—*What are the key tasks? What do they measure? What are the brain regions involved? How can you test animals?*

None of this was part of the working knowledge of molecular and cellular neuroscientists thirty years ago. Nowadays, it's very difficult to find a molecular and cellular neuroscientist who doesn't have quite a good understanding of the psychology that's used to test ideas and principles that they use.

They may not do it themselves—not everyone is doing all of these things together in her own lab—but what you find, which is really different now, is that the language is in place, the knowledge is in place, people know what is happening in other areas.

And this has really made a tremendous difference in the last ten, fifteen years in neuroscience. There's a great deal of integration.

HB: I'm going to ask you to talk about NF1 later on, but as you're discussing what's happening in the lab and who's coming up with what idea and who's implementing what, it strikes me that there is a disparate array of different skills that people have to have to make the whole process work, from developing ideas, to diagnostics, to physically implementing things by using devices, to computer modelling and so forth.

I'm guessing that if you're running a lab it's almost like you're the general manager of a sports team and you have to have everyone work together. Is it kind of like that?

AS: Absolutely. And the trick is to have a group that is collaborative enough so that these arrangements arise spontaneously. If you're a psychologist, you help me with studies that I'm doing because my background is molecular biology or physiology; and since I'm a physiologist, I will help you with your physiology experiments. There's a natural organic that evolves in a lab, but it takes some work to establish that initially.

I remember, many years ago, we had physiologists coming to the lab who were generally motivated to learn some behaviour, but they would come into the lab and essentially they'd just end up doing physiology because they felt uncomfortable doing behaviour. And the same sort of thing happened with psychologists. But over time, when there's this culture of sharing of approaches, this sharing and collaborative atmosphere, then another person comes into the lab and falls right into this well-established tradition: there are rules.

It's not like you're being taken advantage of. You're not losing anything; you're actually gaining something. There are all these unspoken rules that you absorb automatically now, and just the sheer force of how the group acts causes you to become involved. This culture doesn't need to be deliberately orchestrated anymore—the established sociology of the group is responsible for it.

Having said that, of course, as a PI (principal investigator), you need to figure out who should be doing what and who is best suited to do *this* type of work versus *that* type of work.

HB: And new people are regularly entering, I imagine. In my experience, it's a classic sin of academics that they're always trying to recruit people who are exactly like themselves—they're trying to iterate themselves, consciously or unconsciously, all the time. But you can't afford to do that. You clearly have to be thinking globally in terms of your lab.

AS: Absolutely. And what I've learned, what every principal investigator learns, is that you need to have a mixture of people with different talents in the lab: you need people who are incredibly good with experiments, you need people who are collaborative, you need people who are very imaginative, you need all these folks working together.

But often they just happen to come anyway, because there is such diversity in academia that if you just hire people with passion who can get things done, it tends to work out.

Essentially that's what I look for: I look for people who are committed to science who have a history of being able to get things done. Typically, they are all very different at everything else. And this approach does seem to work, actually—it's quite amazing how well it works.

I don't remember the last time that there was a problem in my lab. You'd think with all these Type A personalities working together, competing for a limited number of jobs, you'd think that there would be more attrition, but there isn't.

One of the reasons has to do with the fact that we don't just have one project. We have several projects, so people feel that they have their own arenas and therefore the act of collaboration is not "giving up" something, it's actually "getting" something. That encourages people to work together. In general it's a very positive, reinforcing environment where people feel like they gain from collaboration.

HB: And you were telling me earlier that in your experience UCLA is a particularly welcoming environment.

AS: Exactly—it must be the warm weather. People are more relaxed. It's California, after all. It's LA.

Frankly, I don't know what it is. I joke about it, but it's one of the most important things for a neuroscience community. I'm not sure about other communities. I haven't worked on cancer for many years. At one point I worked in a lab that did a lot of cancer research, so

I've had some exposure to that, but I really don't know much about how it works in other fields.

But what I *do* know is that, for neuroscience, where problems often span molecules and physiology and systems and behaviour and clinical trials, as they have in some of our projects, you can't get away with not being collegial, not being open, not being collaborative.

Otherwise, you are cut out from that which is most exciting in neuroscience today—this integration, this ability to go from a discovery in flies to a clinical trial in humans, from a discovery in mice to testing these findings in functional imaging studies where patients come in and you're scanning their brains.

It's really amazing, but you can only get there if you have a community that's supportive and open. Because if you ask, "*What am I gaining out of this?*" you're often not really gaining a whole lot in the short term, but in the long term you definitely are.

It's tremendous fun to work with colleagues, getting to share all these discoveries. Really, that's what fuels most of us here at UCLA. When you really get down to it, it's not co-authorship—because by the time your name is in the middle of a paper somewhere, you don't get that much out of it anymore—it's just great fun working with people on these issues that are changing mankind, essentially.

Questions for Discussion:

1. As our knowledge of neuroscience expands, will deliberate efforts need to be made to ensure that young people are aware of all key results and core principles in its different sub-fields, or is this something that will simply happen naturally?

2. Are successful principal investigators of multi-disciplinary research labs born or made? Can you "teach someone" to become a better director of a laboratory?

IV. A Leg to Stand On

Understanding changing synaptic weights

HB: Let's talk about the science in more detail now, I've held you off for a little bit, because I do think it's important to get aspects of the back-story. When I'm looking at this, when I'm reading review articles and interviews and so forth in advance of meeting you, I keep asking my self, "*How are these guys actually doing all this stuff?*" and I imagine that other non-experts like me feel the same way.

But it's also clear that the scientific progress in these areas, in terms of pace and breadth, is truly remarkable.

So, I'd like to talk now about memory and learning. You do a lot of different work with memory. Let me ask you a specific question to start with.

I read this piece in *Discover* magazine a while ago about some of your work. The basic thrust, if I remember correctly—ironically enough—is that the common paradigm for memory started from the work of this guy Lashley in the 1950s. His idea was that memory is distributed, so if you have a memory, it's not actually localized in any one place in the brain. And then people started doing some more research and recognized that, "*No, you **can** actually have some sort of a clear, localized sense of a memory.*" And then you took that quite a bit further still and are now able to localize memories almost to the neuronal level.

So I'd like to know where my memories actually are. Where are they? What's our best understanding of where our memories physically are in our brains?

AS: I guess that question is tightly connected to *what* memories are. Maybe I should address that one first, because that will make sense then in terms of localization.

I think your question is quite important, because it reveals one aspect of memory that's really new: our ability to localize memory to the very synapses, actually, to the very contacts between cells that change when you're learning something.

HB: But as you point out, logically there's this first question about memory that we need to address even before that: *What is it? What is memory?*

AS: Yes. Honestly, we haven't known this with great certainty until very recently. Now, some ideas about what memory is originated many years ago, even hundreds of years ago.

Some of the more prominent, useful ideas date back to the turn of the 20th century, when Santiago Ramón y Cajal, for example, suggested that during learning there are changes in synapses—points of contact between two brain cells—and that these changes are then needed for memory. Because if we learn something, it needs to be stored somewhere. And the obvious question is, *Where is it stored? What change in the brain records this information?*

When you store something in a compact disc, you make a hole in the compact disc. When you store it in a hard disc, you change the hard disc. So when you store something in the brain, you need to change the brain somewhat. And more than a hundred years ago now, Ramón y Cajal suggested that this change lay in the points of contact between cells.

That idea was elaborated upon often in the last hundred years, but it has only been in the last five to seven years that we have amassed enough evidence to understand that, during learning, we change the synaptic weights. This is how cells communicate with each other—a greater synaptic weight means that one cell is better at making the other fire.

These changes in synaptic weights are orchestrated by hundreds of molecules, and these molecules regulate these changes in cell-cell communication in the brain, which, in turn, regulates learning and memory. Molecules that trigger these changes are involved in learning, molecules that maintain these changes are involved in memory.

Now, for the first time in human history, we have at hand the sketch of what happens in the brain when we learn and remember something. It's an embryonic sketch: we don't know everything that goes on in the brain when we learn and remember. But now we *do* know one thing. And in science, it's critical to know one thing with certainty, because then we can stand on that one thing and find others—if everywhere we go it's smushy and uncertain, we keep wobbling around and never get anywhere. But now we have a leg to stand on.

To my mind this is probably one of the greatest achievements of modern science, actually: for the first time, we have a molecular and cellular biology of learning and memory.

What are the molecular processes? What are the cellular processes in the brain that are mediating memory? We don't know all of them, that's for sure—there is much to be discovered—but we at least know one general class of them. And to me, that's a **very** big deal, and I'm just lucky that I happened to be there as this was all being discovered.

HB: I want to ask more questions about those specific molecular processes, but, before I do, I'm going to back up and play the sceptic.

So the sceptic says to himself, *"Here's Silva, he's telling me he's been able to isolate, at the molecular and cellular level, the processes that are triggering these synapses: it all has to do with the potentialities, the weights of how strong the signal is, how it's excited, how it's inhibited across the synaptic connection."*

And then he might think, *"Well, okay, but these synapses are just part of larger systems. What I **really** care about are these larger systems that are actually present in the brain: where they are, how these systems work. Because the stuff that can happen at a micro level might be completely different at some level from the stuff that can happen at a macro level."*

How would you respond to that? How would you respond to the claim that, *"Yes, this is worthwhile, this is important, but for me as a systems guy, the **really** important thing is to isolate these bigger systems that are actually composed of these things."*

AS: Well, you are right about this, and I'll tell you why: the most interesting place is the place we have *not* been.

Ten to twenty years ago, the frontier was to come up with unequivocal evidence that the machinery that regulates synapses is an integral part of the machinery that regulates learning and memory.

And now we've done that. So now, the next frontier is exactly what you're talking about: to use these components to understand the complexity of many neurons in multiple-brain regions that interact and allow us to remember this conversation, enable us to remember our family, and all of those things that make us human.

HB: So this is "the leg to stand on", that you referred to a moment ago. This is "the core".

AS: Exactly. Once you know something, there is an implicit simplicity that comes to it.

It's the old story of Columbus standing the egg on its end. He was trying to sell navigation to South America, to the Spanish Queen Isabella. When she asked him, *"Why should I believe in you?"* he replied, *"Well, why don't you ask me something that would prove otherwise?"*

She was having breakfast at the time, so she gave him an egg and told him, *"If you can stand this egg on its end, I believe that you're smart enough to take us there."*

He took the egg, and immediately broke just a little bit at the end—and there it was, standing up! Once you see the egg standing up, then there's no problem: the solution is apparent.

Similarly, once we found out that there were these synapses that change in the brain, that this was an integral and central component of what goes on in the brain when we learn and remember, then that became transparent.

Now what we are looking at is, *How do we take advantage of these mechanisms to regulate the complex cyclic processes, the complex system-wide processes, that allow us to learn and remember?*

And what's exciting about this is just as we were solving the first problem—*What are the fundamental processes of learning and memory?*—a bunch of technology came into being that allowed us to tackle this question of, *What are the system's mechanisms?*

By "system mechanisms", I mean, *Let's not talk just about a single brain cell talking with another brain cell, let's talk about thousands of them; and let's talk about ways that we can understand how these thousands of cells are changing as we are having this conversation.*

How does this evolve into consciousness? How does this evolve into emotion? These are the really deep questions that we've been asking ourselves for probably as long as man has been around. And I think that now, finally, we are in the position to actually ask them with the prospect of making genuine progress.

So I share with you that enthusiasm for the future, which hardly diminishes my enthusiasm for what has been accomplished. In terms of really big accomplishments in science, knowing the molecular and cellular processes of memory has been one of our big goals for so long—and we're finally there.

Questions for Discussion:

1. To what extent might the framework of synaptic weights Alcino describes serve as a basis for the claim that learning and memory should not be regarded as two separate things, but merely different aspects of the same "spectrum"?

2. Alcino mentions the possibility that the cellular and molecular processes mediating memory and learning might be of one general class. How do we know when our list of possible classes is compete? Might there be some sort of meta-framework linking different classes?

*3. Could there be some large-scale systems that should **not** be properly understood as the product of underlying cell-biological mechanisms?*

V. Justified Confidence

How to know that you know something

HB: OK, let's get back to that. I interrupted you earlier to play the sceptic and ask you my question about larger systems, which you convincingly and congenially dealt with. But tell me now about the specifics of these processes, tell me why we're so confident that we have this leg to stand on now.

AS: Well, you'd be surprised. If you were to visit, let's say, ten neuroscientists who have labs like I do, who work on learning and memory, and you'd say to them, "*I spoke with this Alcino Silva guy, and he was all enthusiastic and certain that now we have at hand one molecular and cellular component of what memory is all about*", some of them might reply, "**What?!** *Did he* **actually** *say that? Did you guys go drinking before your conversation?*"

So, that's how new it is. It's so new that there isn't general agreement. You asked the question, *How do we know?* and that is central now in neuroscience, because we have such a large database that it's important to know how we know—*How can we tell when a project gets to a stage that it's essentially completed?*

You would think, from an outsider's perspective, that we know this. After all, we're doing all these complicated things. Of course, scientists must know when things are done and just move on, you might think.

No way: we have *no idea* when things are done. There's a very hot debate on, *Where are the landmarks that are recognizable when you have actually demonstrated something that's unequivocal?*

In science, this is usually done by a question of general agreement, without any clear rules of when we have fulfilled the requirements—it's really amazing.

This is an issue that I have great interest in, and I've been working on this problem, *When do we know that we're done?* for about twenty years.

HB: This is your philosophical, epistemological side that we spoke about earlier.

AS: Yes, I confess: that's my epistemological side. But I hope my colleagues won't hold that against me. I do serious science as well.

HB: Absolutely. So, convince me—I'm a sceptical guy, after all.

AS: OK, so in all seriousness now, what I've done is to try to articulate implicit or explicit rules, because there are two different levels amongst my colleagues. Some of them will immediately say, "*Of course that's what we do,*" while others will say, "*Well, I'm not so sure.*"

The fact of the matter is that, embedded in the way we do science, the way we argue for grants, the way we argue that our papers are right and all of that, there are certain epistemological rules. And one of the things that I have done is to try to make them explicit, not just implicit.

So, how do we know that the machinery that regulates synaptic changes is central to the machinery that regulates learning? It turns out, at least in biology, and neuroscience in particular—I'm not so sure about chemistry or physics—that when investigating a possible causal link connecting A and B—here, "A" being synaptic changes and "B" being memory—there are four main strategies that are typically used. They may not be the only ones, but they are used incredibly frequently. And when all four of these strategies are used and they are repeatedly convergent—meaning that they tell you one story, not ten stories—then you know that you have understood this.

So, what are these strategies?

Well, we can manipulate A—either increase or decrease the probability of A (increase or decrease the levels of A). Those are two strategies: you increase or decrease A.

Another one is that you look and see if A and B covary independent of any direct manipulation.

And the last one is to make sure that you find something else that's important for how A causes B.

And that's basically it.

There are other approaches: for example, in systems neuroscience, what you do is to find out as much about A and B as you can and then model how A interacts with B. Because, classically, in systems neuroscience, we have not been able to manipulate A or B as much, because A and B may be many things spread out through the whole brain and it's hard to get to them.

But in biology, that's basically what you do: you increase and decrease A, you see how A and B covary and you see what mediates between them.

So how do we know that we've gotten somewhere? When you do all four of these things many times, and they just keep telling you the same story over and over again.

You get to a point where this has happened hundreds of times—literally, hundreds of times—and you say, "Well, this must be true."

It's like having an event of some kind, together with witnesses to that event. How do you know that you understand the event? You understand the event when all of the witnesses who witnessed it from different perspectives tell you a single story.

Whatever is in common with all that has been said, you trust. Things that are unique and strange about individual accounts, you doubt. But if the event is important enough, and enough people witness it, you may get an account that's in common amongst all these people, and that's the one you trust.

For us, it's exactly the same. The different types of experiment give us these different perspectives.

Let's say that you take away a gene that you think you need for changes between cells. And now these changes don't take place. And then you look at memory and you see that memory didn't take place either. Can you really say that these changes are critical for memory?

Not really. What about if you took the gene away and something else got changed that you don't know about? That thing could very well be the cause for the changes in memory, not the change in synaptic function.

Similarly, let's say that you increase the levels in a gene and you get better synaptic changes and you get better memory. Can you really say that that's how memory happens?

Maybe by increasing the level of that gene, you just made some thing that has nothing to do with what normally goes on, and now you have better memory, but just like you get better memory or become more attentive when you have a lot of caffeine, it's still not clear what the causal factor is. And so on.

If you see two things correlated, do you know that one causes the other? Of course not. They may have something else that causes them both.

But when you get the same picture with all of these different strategies, and you get it over and over and over again, then you're certain.

What do we have now that we didn't have 20 years ago? We have hundreds of genes that have been manipulated, hundreds of proteins that have been manipulated. They affect synaptic function, and they affect learning and memory.

One of my favourite accidental discoveries involved a meta-study, a study of many other studies. We were just astounded when we looked at all of the studies involving "smart mice".

Believe it or not, there's a bunch of mice that were genetically engineered; and now these mice are smarter than other mice. It's really interesting.

And then we ask in this meta-study, *When we look at all of these "smart mice", when we look in their brains, is there something consistent about the physiology of their brains?*

If indeed, these changes in cell-cell communications are at the heart of memory, then maybe we should see an enhancement in these cell-cell communications.

And the amazing thing is that we found this was true *to a single mouse*—meaning that, of all of the "smart mice" that were in the literature, the ones that had genetic mutations that made them solve tasks faster, essentially *all of them* also had enhancements in the very mechanism that we now know to be at the heart of learning and memory.

And this was done by dozens of different labs. So there wasn't some sort of conspiracy that could bias the data so now all the smart mice have just this type of change.

When I asked myself before I started this meta-study, What did I expect to find? I thought that only about 30% of the smart mice would have these enhancements in these mechanisms that regulate cell-cell interactions. Just 30% or so.

But it wasn't. It was virtually 100%. And that was a huge surprise to all of us. Nobody expected this.

But this just gives you a sense of the kinds of data that we have that allow me to sit here and tell you with a straight face that we understand something truly important about how we learn and remember.

HB: I want to know more about these smart mice, and I want to know more about our current understanding of memory, because I allowed you to drift off from that as well—the problem is that you keep introducing all these fascinating ideas, so I'm not sure if I'm ever going to feel like I'm done.

But let me first pick up on what you've just said and ask a question related to logical structure. You've convinced me that this mechanism of cell weights is necessary for learning and memory—that you can't have learning and memory without it. But what about sufficiency? Might there be something more going on as well?

AS: You are right on track. Remember: memory and learning are at the heart of evolution. Evolution is about shaping creatures that can face the challenge and change their behaviour so that they're better equipped to face that challenge again. That's the heart of evolution.

So it's not surprising that 70% of all genes are expressed in the brain, and probably a majority of those are involved in memory.

This is the thing that evolution has shaped us to do.

HB: Because it's so important.

AS. Right. The problem that organisms have is that the world is ever-changing. When I was a graduate student, we hardly used computers. Just listen to that: "hardly used computers". And I'm not that old—it's not like this was in the 1800s! I first started using *email* when I was a PI (principal investigator) and had my own lab. That's when email came into being.

Anyway, the world keeps on changing. Change has been happening at breakneck speed, and evolution is all about the question, *Can we adapt?*

Our ability to adapt depends on our ability to learn, remember and change. So it's hardly surprising that most of evolution has been geared towards fine-tuning this machine that allows us to survive.

We literally have hundreds of proteins that are involved in memory. I talked about synaptic changes—changes in the communications between cells. But there's more that we've discovered. We don't yet have the same level of certainty for those, but we think that they're very important.

For example, cells can have certain degrees of excitability. If I'm a cell and you're a cell, and you send a signal to me, most of the time I don't listen, actually. Cells are just like teenagers—they listen if you repeatedly tell them the same thing, but most of the time they don't. That's how cells in the brain are.

But if you have a cell that is more excitable, if my excitability goes up because I have just been involved in a memory event, then I will be far more likely to listen than otherwise.

HB: So there is some triggering mechanism…

AS: Yes. We know that there are processes that are constantly balancing things in the brain. If I'm a cell, and I'm involved in learning, there are all these synaptic changes taking place around me. And there are mechanisms that tone down my ability to be engaged so that I am not at the centre of everything that happens, because if I go away—if some specific cell happens to die—then there would be a big hit to the brain.

So you want to distribute information in a circuit. If it's spatial information, it's in the hippocampus; if it's emotion, in the amygdala. You don't want one small group of cells to be engaged in an inordinate amount of memory. So the brain has mechanisms that balance things out, so-called homeostatic mechanisms.

So I don't want to give you the impression that it's *all* about changes in synaptic connections. There's far more than that. But the one thing for which we now have overwhelming evidence, is that this machinery that regulates these synapses, these points of contact between brain cells, is really central to the whole process of learning and remembering.

Questions for Discussion:

1. Why do you think Alcino says "I"m not so sure about chemistry or physics" when he outlines his strategy for biological knowledge? What is it about chemistry or physics that might make them different in this respect?

2. How can the homeostatic mechanisms Alcino mentions be viewed within a context of reductionism vs. large-scale structure? To what extent can we say that structural redundancies associated with evolution provide "large-scale efficiency through small-scale inefficiencies"?

VI. Smart Mice

Objectively evaluating learning and memory

HB: OK, very good. Let me return to the "smart mice" for a moment. What makes these mice "smart", exactly? What have you done to these guys?

AS: It turns out that there is a number of mutations that target these mechanisms of memory. You have proteins that promote memory—promote these changes in cell-cell communication we were talking about earlier—and proteins that inhibit them.

In any complex system, you always have red lights and green lights, accelerators and brakes. It's part of the regulation of complex systems, and memory is probably one of the most complex systems we know. The same feature naturally applies to memory: there are things that promote it and things that inhibit it.

Too much is not always a good thing: remembering everything in this room would be of no use to you, while remembering the right things in this room may be of some use. In creating these smart mice, researchers slowed down the breaks on memory and enhanced the accelerators of memory, if you will.

Both types of things—taking the brakes off slightly and pressing the accelerators slightly—can result in mice that are smarter. Now, why didn't evolution make those mice smarter to start with? That's a more complex question, to which we really don't know the answer. We don't know the cost that those "smart mutations" would have required, had those animals been out there competing in the world of nature.

HB: Is there a limit to how smart you can make the mouse—or a human, for that matter, but I'm getting ahead of myself?

AS: I definitely think there are limits, and the limits are the physical constraints of the brain. You know, "smart" takes into consideration a lot of things: How many things can you hold in mind? How many things can you remember? How many things can you act upon?

Creativity can be seen figuratively as your ability to navigate a problem space, so how quickly can you navigate it? How quickly can you get to that optimal point in the problem space that allows you to come up with that critical insight to solve this big problem that you have been facing for some time?

There are definitely limitations, and part of the promise of understanding the human brain is that we may be able to find ways around these limitations. I mean, we had limitations in transportation, but then we found airplanes, right? So, it wouldn't surprise me if we will continue to develop prosthetics that allow us to deal with the increasing complexity of the world around us.

Our smartphones are such prosthetics—we call on them to remember things that we don't. We call on them to remind us of appointments that we would otherwise miss. We've developed these prosthetics already, and I think that they'll become more and more integrated in our daily lives.

But back to smart mice: I think if we understood how to tinker precisely with the biochemistry, the molecular biology, the cell physiology, of memory, we may be able to call upon it when we most need it. How to precisely do that, that's a different question.

HB: In terms of behaviour, in terms of what we even mean by "smart mice," just tell me a little bit about this. My understanding is that you have this experimental apparatus with mice who can go down various different alleys and remember—or not—where the cheese is or isn't, or something like that. Is that right? What do you mean, in terms of what we're measuring, when you say, "*This mouse is smarter than it used to be*" or "*This mouse is smarter than that mouse over there*"?

AS: Well, just like we have ways of testing how smart *we* are objectively, we also have ways to test how "smart" mice are. In terms of mice, as for us, it means how well they do in different domains.

Let me describe one, which is relevant to both humans and mice. If you come to LA and have never been here, I could set up a task to see how quickly you learn the lay of the land and how quickly you navigate to one of the landmarks that you've been exposed to.

Some of us are very good at spatial navigation—learning a place, finding a way quickly to places we've been. Some of us are not as good at it, we get lost more easily. Aging for example, typically affects that. As we age, we don't spatially navigate as well.

Mice are just the same. Mice can perform "spatial" tasks and navigate in environments they're exposed to, and different ones will master these tasks to different degrees. Just like us, as mice age, they typically get worse at spatial navigation too—which is really quite interesting in itself—but anyway, that's just one example.

Another example is the ability to remember objects. If I show you now any two objects, you'll probably remember them. If tomorrow I came to you and showed you one of those together with a different one and asked, "*Which of these two objects did I show you yesterday?*" you'd probably be able to identify the one I showed you today, but maybe not.

Perhaps if I just picked them up and started playing with them nervously, you might not really pay attention to them—and if I were to ask you about those objects tomorrow, you may have difficulty recalling which ones I played with and which ones were on the shelf that I didn't play with.

That's actually a task that we give mice. We expose animals to two different objects, and then the next day we expose them to two others. How do we know whether they remember? Because mice are naturally curious: they spend more time with a novel object. Why? Well, because if it's novel, it may be of interest: they may be able to eat it, they may be able to do something with it.

HB: So you can gauge that, quantitatively?

AS: Yes: I can tell. For example, an animal may spend 60% of its time with a new object versus 40% with an old object, or 70% versus 30% or 80% versus 20%. And this preference tells us about its memory. Of course, by itself, this wouldn't be sufficient, because maybe some animals like this object versus that one, but there are ways to control for that.

Sometimes, animals have natural preferences, but we can take that into consideration when we are judging memory—just as we can judge memory in humans, we can judge memory in mice. And just as this process is problematic in humans, it's also problematic in mice.

In humans, the process is fraught with cultural and social problems. In mice, there are other problems. Memory is something that you cannot see directly, you cannot look into someone's eyes and see how well they remember. All you can do, with both mice and people, is ask them different types of things and then see how they react. Of course, if they fail, it could be memory, but it could be many other things unrelated to memory.

But when we get a consistent picture, with multiple tasks and multiple trials, we can say with some confidence that these animals do learn and remember better than others.

And when we look across many different mice that repeatedly demonstrate faster learning and better memories, and we find that all of them share one specific physiological signature, then we start to become more confident that we have our hands on something that's really important, we really have our hands on a fundamental aspect of memory and its physiological expressions.

Questions for Discussion:

*1. Are you surprised at the idea that a smartphone can be regarded as a "prosthetic"? For an additional perspective on the continually diminishing line between "the environment" and our brains, see the Ideas Roadshow conversation **Minds and Machines** with Duke University neuroscientist Miguel Nicolelis.*

*2. How might our notions of "intelligence"—what it is and to what extent it is innate—be influenced by our current sociocultural beliefs? Additional perspectives on the nature of human intelligence can be found in the Ideas Roadshow conversations **Investigating Intelligence** with University of Cambridge neuroscientist John Duncan and **Mindsets: Growing Your Brain** with Stanford University psychologist Carol Dweck.*

VII. Manipulating Memories

Turning them off and on

HB: My understanding is that you've actually been able to manipulate very well-defined memories with mice, that you've actually been able to *eliminate* specific memories for mice at doing specific tasks. Is that correct?

AS: Yes. And that takes us to a subject that we were just talking about a little while ago. This part of the conversation started by you saying, "*Lashley saw memory everywhere, but your studies brought memory to specific cells and specific brain regions.*"

One of the big problems in memory is that it's not just remembering. As I said already, there's no point in you knowing all the books on some shelf. You've looked at them, and somewhere in your brain there are already the titles of many of these books, but you're never going to remember them because it's utterly useless to you.

So one important thing about memory is to hone in on things that are important. Another very important aspect is to connect things, because memories are hardly entities by themselves: they are entities on strings, series of events—one event being important to the other.

For example, suppose I were to pause for a second and ask you about an important event in your life, like your graduation from college. If you closed your eyes and thought for a second about your graduation from college, within seconds another event would come to mind. And if you were to hone in on that event for a few seconds, another one would come to mind.

It's highly adaptive. These memories are not isolated: they are part of strings, because, often, one has to do with the other.

If I'm a mouse and I come into this new area where there is all this new food, I would want to remember what brought me there, the time of year that I found this new food, and everything else that happened around it, because this may be information that would help me to predict where I'm likely to find food in the future.

Now, what does this have to do with the placing of memory?

What we've been able to do is to figure out some of the mechanisms that determine which cells are encoding which memory and why—and by manipulating these cells, we can, ahead of time, bias the memory to those cells.

So we understand what gets information where in the brain, and by understanding that we can change the brain so that particular information goes to specific cells in the brain. Now, if we have methods of inactivating those cells, then we have methods of inactivating very specific memories. And that's what we have done.

HB: The only reason I find any of this *remotely* believable is because you've done it. Quite frankly, if we were having this conversation and I would have no particular faith that you've **actually done** this, I'd just say that you were crazy.

AS: That's exactly right. You know, the difference between being crazy and being a scientist is often very small. Actually, you know what the difference is? The difference is publications—because we have this stuff published and it demonstrates that we've done it. But some of this stuff, even to me, sounds crazy.

HB: You were talking before about this idea of thinking about your graduation that makes you think of another related memory, and then another—this idea of triggering memories.

We've all experienced this subjectively: we focus on something and then all of the sudden different memories come to us; or we smell something that we haven't smelled in a while and we get this flood of memories coming back to us. And the obvious question is, *Where were these memories before we triggered them?* Or, maybe, **How** were they there?

But the idea that you're actually able, at some concrete level—at least for mice, in some specific areas—to isolate that with such precision, is completely astounding to me.

AS: Well, I definitely think that this demonstrates that we have learned something about memory, to the point that we can manipulate it in very precise ways.

Now, I may be overselling this a bit, I must tell you, and I should back up and clarify things a bit. We cannot locate *any* memory. I never promised that, and I just want to make sure I'm clear here. I don't want people to think that I'm saying that we've now arrived at something like that movie, *Eternal Sunshine of the Spotless Mind*, where a key part of the plot was the ability to delete memories that may be traumatic or no longer useful to us.

We are not at the point where we can go into a mouse and target any memory that we've given it. But what we can do—which is actually a step in that direction and illustrates that we understand how these memories are being formed, how they are being allocated to specific cells but not others—is change those cells and thus dramatically change the probability that they will be involved in memory.

Later on, we can inactivate them, we can get them out of the picture, once we ask the animal to recall them. That way we can determine ahead of time where those memories will end up that we are giving the animals, and then we can turn off precisely those cells and inactivate those memories, but not others.

Actually, in one of our experiments, which is a favourite of mine, we have given the animals two memories: the first one, that a salty substance was not so good because it made them slightly sick, and the second one that there was a tone that was to be avoided because when they heard that tone, they got a buzz, a slight shock.

What we did then was to change the physiology of the brain in such a way that we determined where one of these two memories went to in the brain, but not the other. We let the other just go into the brain normally.

But for one of them, we funnelled it to a subset of cells in a structure of the brain called the amygdala, which handles emotions. The amygdala cares whether a food makes you slightly nauseous, or whether you want to avoid a tone because you perceive it as dangerous.

And what we were able to do was to get rid of one memory, but not the other. We can manipulate memories in animals by selectively inactivating one memory and not the other. Then we let the animal recover and that memory comes back again. We can literally turn the switch on and off on memories now with the special tools that we have designed.

HB: So, how long did it take for the memory to come back?

AS: It depends on how you inactivate them. Because part of this is that you funnel memories into a subset of neurons, but the other part is that you then turn those neurons off and on again.

There are ways to turn those neurons off for hours, and there are ways in which they are off just for minutes literally or seconds even. We can use light now to inactive neurons, so if those neurons have the memory that we are testing the animals on, when the light is on the animals can't remember. Then the light goes off and the memory comes back.

We can also use light to do the opposite: to turn those neurons on. What we can do—and have done—is turn on the light, and the animals then recall.

How do we know that the animals recall? Because they act as if they are seeing that stimulus that they had already learned about. Then we turn off the light and it's gone.

HB: Let's back up for a moment. My understanding is that you take this memory and somehow you channel it into a set of neurons, which you can then manipulate. I'm okay, more or less, with manipulating the neurons—whether you're doing it with light, or whatever—this strikes me as being in the realm of the conceivable, the feasible.

But when it comes to taking that memory and channelling it into a specific set of neurons to begin with, I'm completely lost. So please describe that process to me.

AS: Very good. This process has to do with the process of how we link to memories, so let me explain our understanding of that process, and then I can tell you how we funnel memories to a subset of neurons.

Let's say that when you got here earlier today at UCLA, you looked at the Gonda building and said to yourself, *"Oh, that's the place where the Silva lab is."* Then a few hours later, we actually met. Now you have the memory of the building and our conversation linked in your mind. How are these two events linked? What happened that linked them?

What we now know from a lot of convergent evidence—well, we don't know this about people, we know this about mice, and we extrapolate from mice to people—is that, when you saw the building, a subset of neurons in your hippocampus—which is a part of the brain that cares about landmarks and spatial navigation—was activated.

When they were activated, they made a molecule called CREB, which directs the synthesis of other molecules in the brain. It's a protein that directs what we call transcription: genes express RNA, RNA is then turned into proteins.

This molecule tells the nucleus of the cell which genes to express; and one of those sets of genes gets turned into RNA and then into proteins that increase the excitability of these neurons—we spoke briefly about excitability earlier: it makes them easy to fire, they need less stimulation to come online.

So now you have a set of neurons that has the memory of this building, and these neurons will fire more easily than others. Then you walk into my office and we start talking, and another subset of neurons in your hippocampus becomes activated, because your hippocampus not only cares about spatial things, but also cares about memories that you recall consciously. This conversation has a lot of semantic content, a lot of episodic content—episodes are like little movies we take of the world.

At any rate, your hippocampus is being activated again through our conversation, and those neurons that are easy to activate will be engaged in storing information about our conversation.

So now you have those same neurons—

HB: We can say that they were primed, at some level?

AS: Exactly right: they were primed. The first memory primed the neurons to be receptive to the second memory. Now you have two memories, and they have a set of neurons in common.

What does this mean to you? In the future, if you think of our conversation, you will think of the Gonda building. You won't think of buildings at Stanford or MIT where you were before, having conversations on different subjects. You won't confuse the two.

It's adaptive to know that this conversation took place in the Gonda building: in case you need this information, it's more accessible to you, as opposed to being forced to link it across the many buildings you've visited in your lifetime and the thousands of conversations that you've had.

HB: These neurons we're talking about are pretty clearly localized, right? They're in the hippocampus.

AS: Yes, they are.

HB: But getting back to a point we were discussing earlier about localization vs. distribution, my recollection is that this guy Lashley was going through and slicing here and slicing there and it didn't seem to make any difference where things were located—the claim was that memories are distributed throughout the brain. But you're telling me that this is happening in the hippocampus. So what's going on? Were his experiments wrong?

AS: No, his experiments were right on. It's the interpretation that was a little misled by the type of experiments that he was doing. Lashley was asking these rats something that required lots of brain regions. And because it wasn't specialized enough, the animal could use multiple strategies to do well on those tasks. So, when Lashley got rid of one part of the brain, another strategy came up. By the way, that's a general feature of the brain. For the vast majority of things that we learn—because we're not just learning one thing, we're learning many things related to this one thing we're learning—if we have a slight handicap for one thing, another part of the brain makes up for it.

When Lashley was making very small, or sometimes larger, lesions in the brain, the animals were able to use other regions to compensate. This was why Lashley developed this idea that memories aren't anywhere, they're everywhere in the brain.

And in a way, this idea is somewhat correct. I was speaking earlier of the hippocampus, a specific brain region. But if you were to use brain imaging to look at what regions in my brain are active when I'm finding my way to this building in the morning, it wouldn't *just* be the hippocampus.

At the same time, I may be thinking about what I'm about to do, about what I did yesterday. I'd be walking, so I'd need control and coordination over my muscles. You'll see my brain active all over the place: my cerebellum in the back of my head because I need to coordinate my movements, the frontal lobes because I will be planning these movements and other things; you'll see many cortical areas active because I'll be looking at landmarks that tell my brain where I am.

There will be a lot of brain activity. And the question, *Which, of all these regions, is critical for my ability to remember where my office is?* is a tricky one to answer, because you have to do lots of controlled experiments to get the answer: no one experiment is going to tell you that.

So, Lashley was at the very beginning of the efforts to understand the brain. He made lesions everywhere, and what he noticed is that it seemed that many of them decreased performance a bit but didn't get rid of the memory entirely. So he concluded, actually somewhat correctly, that the memory's not anywhere.

But what we now know is that there are certain parts of the brain that are *really critical* for specific types of information. Although a lot of the brain may be active, there are certain parts that, if we mess with them, we *really* affect that memory.

For example, we were talking about spatial navigation just now, and I was saying that many areas of the brain are involved in spatial navigation. We know this.

If you go and make small lesions in the hippocampus, our ability to do spatial navigation tasks decreases as dramatically, as it does for mice, rats, monkeys and so on and so forth.

For spatial navigation, it seems that we *really* need this area called the hippocampus. It involves many other brain regions; and if you have lesions in some of these other areas, you also get a relatively small—compared to the hippocampus—decrease in spatial navigation, but when you hit the hippocampus then you get a *big* loss.

The same argument can be made for other areas and other activities. Take working memory—our ability to remember what you've just told me and give an appropriate picture of it, like a sketch pad of what you've just said so that I can begin to formulate a response. For that, the prefrontal cortex is critical. If I have a deficit in the prefontal cortex, I cannot do that. For emotions, the amygdala is critical, for movement, my cerebellum is critical. And so on, and so forth.

HB: So, there's some redundancy, but there are clearly lots of localized areas.

AS: Absolutely—specialization. There is specialization.

Questions for Discussion:

1. How might this chapter help explain what is happening in our brains when we use specific mnemonic techniques, such as a "memory palace" to help us remember thoughts and objects?

2. How might the notion of neuroplasticity influence the balance between specialization and distribution of neurons and their associated networks?

VIII. Individual Differences

Searching for principles in a diverse world

HB: I have a question that is related to something I asked earlier. We've been talking about mice, and we've been talking about extrapolating various arguments and scientific results from mice, at least in principle, to humans. Is there anything that we know that ***doesn't*** map from mice to humans?

AS: There is, absolutely. I mean, the human brain is far more complex than the mouse brain. All we can start to learn from mice are the fundamental principles.

HB: OK, but I can imagine that you have a really complex brain—the human brain—and you have a somewhat less complex brain—the mouse brain—and everything maps over from the mouse brain to the human brain, but there's just a lot more other stuff. It's like an "into map", as you'd say in mathematics.

Or are there things that just ***don't*** map somehow? Could it be the case that, "*This thing here that we learned about the mouse brain simply doesn't correspond to **anything** in the human brain.*"

AS: Answering that question is complicated. And the reason why it's complicated is because there's so much heterogeneity between human brains.

In the news, there are sometimes reports saying that we have invested quite a bit of money, energy and time understanding mice, and treating disorders in mice; but sometimes when we carry out tests for these same factors in clinical trials in humans, it just doesn't work.

Now, this is a real concern—one that I don't share actually, perhaps because I'm not entirely objective on this, since I've spent much of my life working on mice—but it is a fact that some treatments developed in mice have not been shown to work in humans.

From a superficial perspective, then, you could simply say, *"Wait a second, this doesn't work. We're investing so much in mice, and we're not seeing the same treatments working in humans."*

But when you look more carefully into this, what we see is that the process of testing drugs on anything that's brain-related—cognition, emotion, schizophrenia, autism, any of these complex brain issues—is fraught with all sorts of problems. The tasks we do, our measures of whether there is improvement or not, are at this point poor to say the least.

And then you have this important matter of the heterogeneity between humans. We know in psychiatry and neurology, in mental health in general, that drugs that work on one patient don't work on others. The best that we have is that drugs work in about 20-30% of patients.

Ask yourself, *Why is that? Why is it that we have an anti-psychotic that works on this schizophrenic incredibly well, so that all the abnormal behaviour goes away when he takes it for a few weeks, while the same anti-psychotic does not work on 80% of the other patients? Why?*

There are complex answers to this question, one of which is the heterogeneity of the human brain. You and I share many genes that are in common, but we also share many genes that are different. If you look at the structures of our two brains, they will be different, yours and mine, and these differences will have a big impact on whether you will react well to a treatment and I won't, or whether I will be able to do X and Y and you will be able to do Z and W.

So when we ask the question, *How much of what we learn from animal models—from mice, from rats, from monkeys, from flies from worms—applies to humans?* there isn't an easy, straightforward answer.

One thing we know, which I mentioned earlier in our conversation, is that there's a tremendous evolutionary conservation. That's undeniable. The data are overwhelming: neurons in the C. elegans worms have some of the same properties as neurons in human brains. Some of the same brain regions are there in mice and humans, and many of the genes are identical.

But just like there is heterogeneity between human beings—how they react to drugs and what behaviours they express—there is, of course, a great deal of difference between mice and humans. So it's not surprising that some of the things that we discovered in mice we don't find in humans.

By the way, mice are not just one thing either. We think of them as just one thing, but they're not. Mice are many things, because, just like we have Europeans and Asians and Africans, in mice we have different strains with very big differences genetically.

If we develop a treatment in one strain, it won't necessarily work applied to other strains of mice because of this heterogeneity, this biological richness, which allows species to cope, through evolution, with an ever-changing world. This sometimes goes against us when we are trying to develop a treatment that we so dearly need.

In short, it's hardly surprising that some of the things that work in mice don't work in humans, because most things that work in humans don't work in humans!

Is it the case that, sometimes, mice mislead us? Absolutely. There are differences in the biology. But as scientists, these are not fundamental problems, these actually reflect the richness of the systems that we use. And these differences sometimes highlight what's important.

Right now, we're facing a struggle to optimize the process that takes us from understanding the brain to treating and curing the brain. Do we know what works and when? No. But we already have some things that do work—that I can say with confidence.

Questions for Discussion:

1. To what extent can the degree of effectiveness of a drug in a given population tell us something about the evolutionary priority of the underlying physiological mechanisms the drug is targeting?

2. How might advances in neuroscience help us better address the question of whether or not different species evolve at different rates?

IX. Treating Cognitive Disorders

Towards reversing cognitive deficits

HB: I recently watched a video of a lecture you gave in Australia about a particular treatment dealing with this NF1 gene. The whole story struck me as remarkable and I'd like you to tell it again shortly.

But a fundamental point that I took away is actually something even more general and inspirational: that the causal agents for many neurodevelopmental disorders—the genes or what have you that are responsible for that specific lack of appropriate development that occur when the subject is young and developing—are somehow still present in some form in the adult developed brain. And this, in turn, implies that there's something that might well be done to address the situation, some treatment that could be applied.

That is something that seems quite counterintuitive. You might well think, *"A developmental disorder is a horrible and tragic business that prevents a young brain from developing normally, but once someone is an adult and his brain has fully developed as best it can, there's nothing more that can be done."*

But your message seems to be, *"No, actually, that's simply not true, based upon our research. You* **can**, *in fact, do things—you can in principle develop treatments,"* primarily because, as I understand it, the same agent that was responsible for inhibiting proper development still plays some active role; so it's possible, in principle, to get in there and modulate or somehow change things in a fundamental way.

That overarching message is an obviously inspirational one for anybody who's personally affected by such developmental disorders, but my sense is that it's very optimistic for everyone, suggesting that there might be much that can be done to enhance life on so many different levels, including aging. To me, this is all deeply impressive and unequivocally positive.

AS: Thank you. One of the happiest days of my life was when we first demonstrated that we could reverse the cognitive deficits, the learning and memory deficits, of an animal model of neurofibromatosis type one (NF1) in the adult mouse.

As you so well indicated, there was a tremendous bias prior to 2000 that for this large class of disorders that affect development—schizophrenia, autism, intellectual disability—the genes that cause these conditions act initially during development. When you look in those brains, there's a lot of evidence that they don't look like other brains that don't have those problems.

And there was also this well-understood bias that, if we can't address this problem in development, by the time we have an adult that comes to a medical office and says, *"Help me, I don't learn the same way, I can't pass my classes"*, or a parent comes to us with a child that has obvious intellectual or behavioural problems, it will be too late.

That bias—that it's all about development, that we won't be able to do anything after development—was so deeply ingrained in neuroscience that when we designed experiments to test it we had to design these experiments multiple times because no one was even willing to try.

People thought, *"That's crazy. Why would we do this? This doesn't make any sense."* That first experiment in our lab was actually done by an undergraduate: naive and willing.

We had these animals with a specific deficit, together with a drug that was targeted at the biochemistry of a certain gene that we believed was responsible for mediating those deficits. And the results were unbelievable. We ran this experiment enough times that we became convinced that we could reverse these deficits.

This work was, of course, fuelled by more advanced scientists in the lab: people like Rui Costa, who spent a great deal of time looking at this problem systematically and very carefully. What Rui published in *Nature* in the early 2000s, was that, for neurofibromatosis type one—NF1—we ***can*** intervene in the adult and essentially reverse the learning and memory deficits that this animal showed.

And that was a great day.

Because not only did that have obvious implications for patients—1 in 3000 people worldwide have NF1, with roughly half of them having some degree of learning problems, and what Rui's results demonstrated was that, for the first time, we might be able to intervene and help these people—but these results also demonstrated something that we should have known already: that there's a tremendous amount of flexibility and plasticity in the human brain and we systematically underestimate the brain's resourcefulness at recovering and repairing itself.

The brain is not a Newtonian machine, with one lever triggering another, and another lever triggering another, and so on. The brain is highly dynamic. And if we can just help it to restore some of its functionality—not *all* of it, but *some* of it—the results can be nothing short of magical.

In this case, the result meant that adult animals—***not*** in development, it should be emphasized, we didn't touch them in development—that were unable to learn very well, that had problems with spatial navigation and interaction with their environments compared to other animals, could have their learning deficits reversed when we addressed the biochemistry that was affected with a specific drug.

Then, after our study in *Nature*, several other labs, some of which were doing work more or less in parallel, published dozens of examples like this—neurodevelopmental disorders that we were convinced were *all about development*—nothing to do with the adult brain whatsoever—and then intervening in the adult brain and demonstrating remarkable improvements.

This has been one of the really important things that we found, something with the greatest potential for human impact, because as you said, it opened the door to the possibility of treating literally hundreds of millions of people worldwide with problems that we never thought we would ever have a chance of even touching.

It's a long road, but it's one that we have started upon—not just my lab but many other labs as well—and there are some early, hopeful signs that we may be on the right track.

HB: I'd like to talk a bit about some general structural matters on your approach to neuroscience that I picked up from your talk.

You point out four discrete stages leading towards a possible treatment for some of these conditions.

The first one is to identify which particular gene or causal agent is actually at work. The second one is then to ask, "*What sort of animal models do we actually have to be able to test these ideas?*" The third stage is to identify the molecular causal mechanisms of the condition, while the fourth stage is the investigation of possible treatment to address those mechanisms.

Is this a fairly standard categorization scheme of how to go about addressing cognitive disorders from first principles—these four stages? Is this something that most of your colleagues and peers would agree with?

AS: You know, it really depends on the type of cognitive disorder. Take memory. It's true that one subset of disorders that afflict memory in patients is clearly genetic. Others have a genetic component but they have other components as well, such as environmental factors like alcohol abuse. There are a number of other things that can affect memory that don't really have to do directly or immediately with genes. Genes are playing a role, but not a direct role.

If we talk about genetic disorders as one concrete example, then the hope is that, if we understand what goes wrong when this gene is changed—when this gene is absent, for example—if we understand with clarity the cellular changes, the systems changes, and how they are accounting for the behavioural changes in an animal model, then we can better design treatments.

And these treatments will often involve drugs, but not *just* drugs. Treatments for memory will involve more than just drugs. There are also important behavioural aspects. You can imagine a teenager who is far behind in school, someone who has an IQ equivalent to a seven-year-old, for example. Suppose you can then magically treat the brain mechanisms that are involved in learning and memory. That individual wouldn't instantaneously be a normal teenager, because there is all this learning, all this socialization, which would also need to be addressed.

We are such sophisticated animals and we are learning constantly—our social interactions, our knowledge of the world, our knowledge of what is needed to advance through the educational system and become professionals and all of that. There will be much that needs to be done to such individuals to help them catch up.

To make an analogy, imagine a new computer without any software or memory chips. In order for that computer to be equivalent to the computer you're currently using, you'd need to somehow copy all those files, all those programs from the old computer to the new. And it's a lot of work to transfer stuff from one computer to the other.

Although it's not terribly enlightened, in a way we can use that image too for the human brain. Even if you restore the machinery, you haven't restored all the software, all of the memories that allow us to function as independent human beings.

Getting back to your question: are these stages standard? Well, nothing is standard in science: science is about violating standards. But do they capture a general research program? Absolutely. And the research program is, *Let's find the genes that affect and disrupt memory in humans; let's then mimic those genetic changes in cell lines in mice, in rats, in C. elegans, in drosophila—anything that is useful for us to understand those genes, let's figure out what's wrong there— then let's figure out appropriate models to see if we can reverse it as a preamble for what we want to do in humans, which is to bring these drugs to humans.*

It should be said that there's often a price to use these drugs— there's no such thing as a safe drug. Look at aspirin, even. People die from aspirin all the time. So, there's no such thing as a perfectly safe drug: safety is a relative thing.

And before we can bring this into human trials, before we can give this to our children, we need to get a sense of how safe these treatments are. Because we certainly don't want to make things worse—the last thing we want is to take a population that's already affected and make things worse. That's why one of these stages you mentioned is to try these things out on animals. We don't want to hurt animals, but we'd rather hurt animals than our children. And once the treatment is ready, once it's mature enough, then we bring it into clinical trials so we can test its efficacy with patients.

Are there variants on this? Of course—as many as there are scientists. But is this a common core of what many labs are trying to do? Absolutely.

HB: Another aspect of your talk I found intriguing was the precarious sense of balancing between inhibitory and excitatory activity in the brain.

For those who had a cognitive deficit that was associated with an imbalance between the two—they had more inhibition or more excitation than they should have had—the specific mechanism to remedy this might work very well for them, but for people who were already in balance, it could be very detrimental.

To me that demonstrates two things: that you have to be extremely careful about how you would give any proposed medication, because unless these people suffer from exactly that particular condition that would render them susceptible to that precise treatment, you might be doing more harm than good.

And it also shows the amount of complexity involved in all these different conditions. There are so many different symptoms and so much complexity that it's clearly not as if there's going to be one miracle drug that will be great for everyone.

AS: That's exactly right. It's not uncommon in medicine that a drug that helps some group of patients actually makes things worse for another subset of patients. Clinicians are very attentive to this possibility as they follow patients in the early stages of development of treatments. Once again it's a display of our heterogeneity: we are very different as human beings, you and I as well as everybody else. So, there's that component.

When we were developing the treatment for NF1 in mice, we went way out of our way to find the treatment that didn't make the normal mice worse: a treatment that helped the mice with the mutation that mimicked NF1, but did not have any negative effects on the normal mice.

Why? Because we reasoned that, if we found such a treatment, then the chances of causing harm once we brought it into humans would be decreased—it wouldn't be eliminated by any means, but it would be decreased—and, indeed, the treatment that we developed in mice, for the most part, *has* that property.

What we found, as you mentioned, was that there was an imbalance between inhibition and excitation: the inhibitory systems in the brain were abnormally high. As a result, the kids with neurofibromatosis—not all of them, some kids are perfectly normal with neurofibromatosis, but about thirty percent or so—had problems with learning. But if we gave these children some additional training, they learn.

In other words, the brakes are on, but if you press the accelerator gently, you can sort of overcome that. The system isn't optimized but you can overcome that.

In mice, it's the same way. The brakes are on a little tighter in mice, but when you work with the mice, they learn, just like with the kids.

And what we found is that the drug that released some of this inhibition, allowing us to teach the mice faster so that they learned just like the normal mice, that same drug at that same concentration did not also cause deficits in most tasks that we asked them to do. In some tasks it did actually, but in the majority of them, it didn't. And that gave us a great deal of comfort that we would not first cause harm as we aimed for a benefit.

By and large, what the clinical trials that followed showed was exactly that: although the jury's still out on whether this drug would work in humans, the trials that we've gotten back so far clearly show that they have not hurt.

Now, is this always the case? Of course not. But in this case, it was.

I wish I could take credit for this discovery but I can't. I still remember vividly, late at night in this very office, when a student of mine, Steve Kushner, came in the door and said, *"I've been doing my rotations in internal medicine, and I've found that this drug, statins, at concentrations that are quite benign, can act to reverse the biochemistry that we've discovered in the lab is behind the learning deficits in the mice."*

Honestly, I did not believe him. And the reason I didn't believe him was because, if this drug, statins—which people take because of high cholesterol—is having an impact on NF1, how many other proteins is it also having an impact on?

That drug could potentially have an impact on 20% of all proteins in the body, so I thought, *"Well, wait a second, if it's really having an impact the way you think, then we would definitely have noticed. Maybe it helps me if I've got NF1, but otherwise you'd be disrupting it. And if it is generally disrupting 20% of all proteins in the body, this drug would probably be the most powerful poison ever known."*

But the magic that I didn't predict was that the drug would reverse the deficit of the mutants at concentrations that did not touch the controls at all.

Had the drug touched the controls, it would have killed all the mice that we tested it on. It would have killed all the humans who take statins, because statins control the levels of these little fats; and these little fats, when added to molecules, allow these so-called "Ras molecules" critical to learning deficits to attach to the membrane.

But what this brilliant student—he is now a professor of psychiatry in Rotterdam at Erasmus MC—realized is that you could get rid of this lipid group by taking statins, causing the Ras molecules to go away from the membrane; and because the problem to start with is too much of this Ras molecule at the membrane, you can now fix the problem.

My problem was that if you are taking these molecules away from the membrane, what about all the *other* proteins that also have the same little fat group? Won't they be affected as well?

But what I didn't realize was that the levels of the drug that take these Ras molecules away from the membrane were not high enough to affect all the other proteins that use these little lipid groups; and the reason for that is that there is too much Ras in the membrane to start with. So, if you decrease these little fat groups, the one that gets affected first is the Ras, the other ones are not terribly affected at first.

HB: So it's a question of scale.

AS: Exactly. And that is what allowed this to work. So thank God I didn't tell Steve, *"This is a horrible, stupid idea. Why are you bugging me? I want to go home, it's 11 o'clock at night!"* Instead—and this is the only credit I can claim—I didn't say any of that. I was completely convinced that there was no way that this was going to work, but he was so enthusiastic that I wasn't going to pop his bubble.

You see, one of the most important things in science is the enthusiasm and force behind an idea. My experience has told me that anytime there is someone as bright as Steve, with all that energy behind an idea, let him go, because something good is going to come out of it.

And as I said, that's the only credit that I can claim for that discovery. Everything else was Steve and Weidong Li. The two of them paired up and showed that you could use statins to reverse the molecular deficits, the cellular deficits, and even more importantly, the behavioural deficits for NF1 mice.

And that was a *good* day.

HB: I can well imagine.

Just a word about NF1 behavioural deficits in humans: you mentioned attention and planning, but I also understand that, for humans, there are sometimes other issues, such as vocabulary.

AS: That's right. The kinds of deficits that we could capture in mice that some NF1 patients have—again, not every patient with NF1 has this, only about 30-40% do—are problems in memory, spatial navigation, attention and motor coordination. By and large, the statins were able to treat all of these problems in mice.

But, of course, there are problems that you see in patients that we couldn't model in mice: reading skills, for example, and vocabulary.

HB: You've mentioned several times now that only about 30% of patients who have NF1 manifest these cognitive deficits. Why is that?

AS: It's called partial penetrance. Every human disease has that.

The reason is that, in addition to the neurofibromatosis gene that causes NF1, there are other genes that interact with it behind the scenes, making the effects better or worse.

HB: So they might compensate in some way.

AS: Exactly. Just now, I told you how there was too much Ras in the membrane in the case of NF1. Imagine that the protein that adds these little fat groups that allow the Ras to be in the membrane doesn't work so well. Normally, you would have too much Ras in the membrane, but because the protein that adds this fat group doesn't work so well, now the levels are normal and you'd learn perfectly fine, despite the fact that you have a mutation in NF1.

HB: You have two things that cancel each other out.

AS: Exactly. And we tested that directly.

One of the most magical experiments that we've done—it was performed by Rui Costa—was with two different mice with two different genetic disorders caused by mutations that impair learning. Both had problems learning. Then we placed both mutations in the same mouse and that animal learned completely normally.

We are the result of all of these biological forces working together, determining whether we can remember, whether we can learn, whether we can pay attention. And often they are in opposition to each other.

HB: These systems are so complex, it's almost a wonder you can get anything done at all—there are so many variables.

AS: That's true. There are many variables. But at any one point evolution worked on one or two of them, and then those allele combinations became dominant because they were so adaptive. Those are the ones that we see today. And those are the mice that we study, because we engineer them that way.

It is true that there is a great deal of diversity. It is also true that, experimentally, we can fix that diversity artificially—we can make animals that have twins. Essentially, the strains of mice that we work with are twins; and then amongst this similarity, we can then see the differences when one of these genes is mutated. But in humans, of course, things are more complicated.

HB: So you generate your own twin data, basically.

AS: Essentially, all of mouse genetics is done on what amounts to twins: hundreds of twins, where we manipulate this gene or that gene and then see what happens. And we do this because this diversity, which is our evolutionary friend to enable us to adapt, works against us when we want to simplify and understand things.

Questions for Discussion:

1. Might the notion of partial penetrance be used to help re-evaluate our very definitions of genetic diversity? Readers interested in the concept of how to rigorously assess genetic diversity are referred to the Ideas Roadshow conversation **Our Human Variability** *with geneticist Stephen Scherer, co-discoverer of the phenomenon of widespread large-scale genetic copy-number variation.*

2. What does Alcino mean, exactly, when he says that "those allele combinations became dominant because they were so adaptive"?

3. Do you think there will be a time when diseases will be primarily defined by genetic and cellular characteristics rather than large-scale behaviours and associated medical susceptibilities? If so, how might this affect our very notion of the concept of "disease"?

XI. Justified Optimism

Making a difference, today and tomorrow

HB: Let's talk about the future now and how things might improve in terms of treatment.

AS: I choose to see the situation from a hopeful perspective. These problems have been there since the beginning of humanity. There are historical accounts, descriptions of these diseases that are as old as the Ancient Egyptians. But finally, we are at a point in history that we understand enough about the brain that we have the hope of being able to intervene—so much so that we have been able to intervene in animal models, and we're starting to get to the early days of studies in humans.

For example, the very same drug that we used in mice—lovastatin, a type of statin—has been used in humans with some positive effects. We don't have the numbers we need yet, but we hope to have them soon.

One of the great things is that these results got other people, other labs, involved in testing the clinical implications of our findings. There are large, ongoing, clinical trials. So in the near future, hopefully, we will know whether or not statins worked in these cases.

For lovastatin specifically—not all statins are the same, of course: they go into the brain at different rates; and as we said, you have to get the right levels: too much Ras can be just as bad as too little—there seems to be some real hope there, based on what we have learned from early, very small, clinical trials.

There is a paper that was just published showing that NF1 patients have enhanced inhibition.

They didn't do the studies like we did—in mice, you can measure directly from these neurons.

But what they were able to do was to stimulate motor neurons by stimulating muscles, and then record responses in the brain. They can test inhibition from these responses, and they found that NF1 patients, just like NF1 mice, have enhanced inhibition.

They also have a way of testing plasticity, synaptic plasticity. It's very indirect—it's an echo of synaptic plasticity, because you can't do it directly in the brain, for obvious reasons—and they were able to show that, in addition to enhanced inhibition—just like we showed in the mice—these patients also have a deficit in plasticity.

HB: How do you quantify a deficit in plasticity?

AS: Well, plasticity can be triggered by training. That's what causes learning. If I tell you about things enough times, eventually you learn them.

Now, with this technology, instead of testing learning—because it's too complicated to follow these responses—what they've done is to stimulate different muscles and record from the motor cortex. They stimulate them in a pattern that's similar to the pattern that normally triggers learning. It's the same arrangement, the same architecture.

And what they've seen is that in normal brains, you see a potentiation—an increase in the signal over time—of these muscle-triggered responses, just like you'd see a direct potentiation of these responses during learning in rodents.

This is all very indirect, but the amazing thing is that they *did* see enhanced inhibition, they *did* see deficits in plasticity; and both were *reversed* by lovastatin treatment, just like they were in mice.

And they also showed that the patients who had problems with attention were improved by treatment with lovastatin as well. Now, I don't put much faith in that because it's a very small trial, but it was just striking, the specificity of these results.

HB: How significant were the improvements relative to the placebo effect?

AS: Well, this was a double-blind, placebo-controlled study. When you don't have a placebo control, the placebo effect is so large, it's hard to know what to make of any of it.

I should tell you, for completeness' sake, that a student of mine carried out a large trial in the Netherlands, not with lovastatin but with simvastatin. When we were thinking about it, the two of us, we thought simvastatin would be much better because it goes through the blood-brain barrier—your brain is separated from the rest of your body by a filter: stuff that's in your blood, in your heart, in your kidney, in your gut doesn't just get to the brain, it first has to get through the blood-brain barrier. So we decided on simvastatin, not lovastatin. Lovastatin is what we used in mice, but simvastatin is what we tried for that trial because, in Europe, simvastatin is approved for use, but lovastatin isn't, and it penetrates the brain much, much better—ten times better, in fact—and we thought that it was going to be much better.

Not true. He ran a sizable clinical trial and it failed.

We don't know why it failed, but I suspect that it failed because too *much* of it got to the brain. And we know that both too little and too much Ras is bad, so maybe too much got to the brain and that's why it failed. We don't know.

But the good news is that there is a very large trial that we've been a part of, and this trial will become "unblind,"—which means that we are going to figure out who's who and what the effects were—in a few months. So hopefully I won't have to wait very long until I have a clear sense of whether these decades of work have actually translated into helping a group of people or not.

HB: Well, come on, you're being too modest. You're obviously on the right track. Getting back to what you said before about having a leg to stand on, you clearly have a strong basis for real progress. Which specific drugs, which specific doses—all of that might be uncertain right now and perhaps for many years yet—but we're obviously at a new frontier.

AS: Yes, I actually think that's exactly right. This is the very beginning of what will be a large number of similar studies in, for example, intellectual disabilities. This is untracked ground: we just don't know what strategies will work, how we should go about tackling this problem.

So even in the cases for which we may not be initially as lucky, the process is still teaching us a lot about what we need to do. I don't think, at this point, if the science is carefully done, that there are any losses: they're all gains, because we've never done this before.

This is one of the biggest challenges of mankind. We've had these problems since the dawn of history, and now we're finally at a time where we're on the threshold of being able to address them.

We're not always going to be successful—of course not. But we are learning how to increase the rate of those successes. And I can't think of anything more exciting to do with one's life right now.

It's very personal, very subjective obviously—everyone in science will say the same thing about anything they do. But I tell you honestly, I feel very lucky to be here right now because there is something truly historical happening.

I'm convinced that our children, certainly the children of our children, will have a very different relationship with these types of disorders than we have with them today. Think about the time before antibiotics, how the world was—a simple cut could kill you. Now, we don't even think about it: if we cut ourselves—even sometimes severely—we just wash it, treat it, take antibiotics, and most of us easily move on.

Just imagine a time when for parents of a child with neuro-fibromatosis, with autism, with schizophrenia, with any of these horrible, horrible disorders, it will be like a cut—something you need to address, something you need to treat, but not a life-changing condition like it is now.

You know, the lifespan of parents with children with intellectual disabilities like autism and schizophrenia is actually *decreased*: that's how severe those disorders are for families that have to deal with them. So, to even be among the pack of individuals who may be able to do something about this keeps me up late at night and gets me into the lab early in the morning, because what else could one spend a lifetime doing that is more worthwhile?

Questions for Discussion:

1. Do you foresee any ethical concerns associated with future medical advances? If so, how do you think they should best be dealt with?

2. If there was a pill you could take to improve your intelligence, would you take it?

XII. Managing Discovery

Harnessing opportunities in an open and mature way

HB: I could literally talk to you for ten more hours, but I won't because you've been extremely generous with your time. I will ask you one more question, however. Suppose you had the opportunity to have any scientific question you wanted answered: imagine that I was some all-knowing guru and I could tell you whatever you'd like to know. What would those questions be?

AS: To me, there is not even a close second. There is one thing that modern biology needs to know more than anything, which is, *How are we going to navigate the immensity of this information that we have generated?*

It's way beyond human dimensions. It has been way beyond human dimensions for a couple of decades already. We have close to 25 million papers published in *PubMed*, probably as many as nearly 300 million experiments published.

Take learning and memory. A conservative estimate is that there are maybe 4–5 million experiments published in learning and memory. How can anyone have even a passing familiarity with this literature so that we can optimize its use?

So if I had a chance of flicking my finger and knowing something, learning something that would make the biggest difference, there is absolutely no doubt that for me what that would be is a strategy, a way of navigating this immensity, so that we can take advantage of it and not be crushed by it. That's what's happening right now—it's mostly crushing us.

HB: And that is now one of your research projects, I understand. Your book, *Engineering the Next Revolution in Neuroscience* (co-authored by Anthony Landreth and John Bickle) is all about this.

AS: Indeed, it's something that I have been limping along with, trying to do something about it, because in my view not much is happening, not fast enough.

There are a number of computer scientists worldwide who are dealing with this "big data" problem in science, but so far the big data problem has really been focused on organizing this long list of genes, organizing these complex patterns of brain activity, and so on. It hasn't really been directed to organizing the immensity of the published literature.

We definitely need all that has been done in neuroinformatics, that field of neuroscience that has organized information about the brain. We need all of that, of course. But I think that more than anything, more urgently than anything, we need a way of navigating the immensity of this literature.

And in a very modest, in almost a childlike way, my colleagues and I have tried to do just that: find a way to organize these experiments, find an interface so that scientists like myself can interact with them as we interact with our Google maps.

Think of the immensity and complexity of Los Angeles. It's 90 kilometers across. To find your way from the airport to Caltech and to other places that you went to in order to have interviews like we're having now, is very complicated, but you had your Google Map and you were able to instruct it to tell you how to get from Pasadena to UCLA and so on.

That's what we need. We need to have all of this information in a place that allows us to query it and interact with it just like we interact with our Google Maps. The complexity is not a problem. The complexity is what helps us: the more we know about the brain, the more we can figure out how to help people, how to answer the questions that we want answered.

The problem is that we haven't progressed enough in terms of developing the tools that allow us to *use* this information. Imagine the world of finance without computers? Imagine if we couldn't make transactions every few seconds, or every few hundred milliseconds, how different the world economy would be.

In a way, that's where we are with neuroscience: we have the same complexity as the world of finance, but we don't have the tools to deal with this complexity in ways that would allow us to navigate it and use it in our favour, instead of against us.

HB: So, the thinking is that you need a sense of structure to see some objective and meaningful overall patterns out there so that researchers can get some sense of promising areas based upon evidential reasoning and clear results without having to feel overwhelmed by 25 gazillion papers out there.

AS: Exactly. We need simplification principles.

I mean, think about it, why aren't the number of roads in the world a problem for Google maps? There are probably trillions, certainly billions, of different, little roads in the world. How come that's not a problem? How come that's an advantage for Google maps?

The reason is because that information is structured. In maps, all roads have a symbol, all highways have a different symbol. So when we look at a map, that's not a problem, that's an advantage that the mapping provides.

Look at the thousands of rivers that are all in Google Maps.

That's not a problem. That's an advantage, because it represents a landmark that we use—the same with oceans, cities, and so on.

We need a similar type of organization that would allow us to intuitively navigate the immensity of scientific results as intuitively as we use Google Maps to navigate and find our way in different cities around the world. That's what we need.

HB: Is this catching on? Is this sense of what is now needed one that is becoming more and more endorsed by your peers?

AS: Absolutely. Big data is big everywhere, including science. I organized a meeting recently at UCLA on this very topic. And I just came back from another meeting on this issue this past weekend.

So there are efforts. They are still embryonic. This is not a problem that has a lot of economic force behind it at the moment: there aren't a lot of labs working on this, but it's definitely a problem that's recognized.

NIH has this *Big Data to Knowledge* program that they've put in place. It's on the order of tens of millions of dollars, it's not on the order of hundreds of millions of dollars. But it's a beginning.

The thing that makes me so anxious about this is that we have the technology to do it. This is not a problem that we can't solve. The only reason we haven't solved it yet is because we haven't yet seen with sufficient clarity that it *is* a problem.

But it's a problem that we can solve. And solving it would have a huge impact on the optimization of science. My graduate students spend 12-16 hours per day with science—there's a tremendous involvement.

Our country, NIH alone, spends 33 billion dollars. We're putting so many resources behind science because we realize how important it is, and I think it's time that we also put significant resources on how to optimize it.

HB: That sounds like a completely reasonable argument to me. Is there anything that we haven't touched on? Is there anything we've left unsaid that you'd like to communicate?

AS: Well, there is actually something, which makes me very uncomfortable to talk about because it's a topic that's usually shied away from.

We have children with horrible genetic disorders, and we want them treated. We have many problems with very serious diseases—cancer, heart disease and so on—that destroy our lives and we badly want them treated too.

But often, we shy away from facing the hard questions of what we need to do to treat them. And one of the things that we are often forced to do in order to find a treatment is to do experiments on animals.

Now, there is a growing movement worldwide against this sort of experimentation in science. I respect those people—profoundly, actually.

I share the questions that they're asking, and I share their concerns. But the movement has become so radicalized, that some of them—not all of them, most are peaceful: they have concerns for animals and they express them—have used terrorist tactics.

I have colleagues who were forced to abandon research on disorders we want treated because their cars were firebombed.

There are real, ethical problems that are difficult and need our society to take a mature, thoughtful look at that we are *not* discussing, that we are *not* talking about, because these questions are being taken over by individuals who are radicals on both sides—scientists who think it's okay to do everything and individuals in the public who think that we can't do anything.

Somewhere there is a very difficult balance, one that I struggle with every day. There's not a single experiment in my lab that gets done lightly. We respect these animals. These are little mice. We love them. But we love our children, and we want to address these heavy and very difficult problems that we discussed in our conversation.

This is a difficult issue, one that we have not come to terms with, because we haven't sat down at the same table and talked about them maturely. We desperately need to have this conversation, we desperately need to educate the public at all ages, so that we have clarity and lucidity about the choices we are making. They are difficult choices, but the stakes are far too high to ignore.

We can't walk away from the people next door to us at UCLA's Ronald Reagan hospital who are literally dying—slow, painful, terrible deaths—without any thought to what's behind the treatments that might help them. We need to face these issues. They are difficult, they are painful, they involve very difficult decisions. But we need to address them head on in mature ways that are helpful and constructive. I really think this needs to be done.

HB: Alcino, I have to tell you: I feel like I've become the president of your fan club. Thank you so much for your time.

AS: It was really a pleasure. Thank you

Questions for Discussion:

1. Are some areas of inquiry better at investigating how to efficiently use their resources than others? To what extent might neuroscience be able to learn from other fields?

2. Might current approaches to "big data" actually distract us from best harnessing current technology to address pivotal structural issues in neuroscience? For additional, and quite different, perspectives on issues related to "big data" interested readers are referred to the Ideas Roadshow conversations **The Social World, Reexamined** *with Tufts University philosopher Brian Epstein and* **The Value of Voice** *with LSE professor of media, communications and social theory Nick Couldry.*

3. Do you agree with Alcino that we need to have an open and honest public conversation about how best to use animal models in medical research? How, practically, do you think that we should go about having such a conversation?

Continuing the Conversation

Readers interested in more details about how we can better manage scientific knowledge in neuroscience are referred to Alcino's book *Engineering the Next Revolution in Neuroscience: The New Science of Experiment Planning.*

A Matter of Energy

Biology From First Principles

A conversation with Nick Lane

Introduction

The Big Picture

While many scientists welcome the opportunity to engage the public with their work, most don't write popular books about it. There are many reasons for this, ranging from the fear of trivializing subtle technical concepts when rendering them in everyday language to a broader philosophical belief that the advocacy of specific scientific views should best be left to designated academic mechanisms like peer-reviewed journals.

For most, however, the major stumbling block to writing popular accounts of their work simply boils down to a question of time, aware as they are that the act of conveying advanced scientific concepts to a general audience requires a tremendous amount of effort to do it properly. Given the many priorities and demands on her time that an active researcher has, the decision to refrain from writing popular books is eminently reasonable, but the natural upshot of this is that the domain of popular science is principally left to a combination of "professional popularizers" who aren't themselves engaged in any real form of frontline research and retired authorities looking to cement their scientific legacy. Both can certainly be interesting and worth reading, but neither is likely to give you a clear sense of what is actually happening in today's laboratories and why.

But then there is UCL's Nick Lane, author of no less than five highly detailed popular books on evolutionary biology and the specifics of energy-conversion mechanisms in biological organisms, who consistently refers to himself as a biochemist *and* a writer. How is it possible that Nick can be such an exception to the rule?

Well, one key factor, it seems, is that he had a different sort of scientific career trajectory from most, one where the act of writing played a significant role for him relatively early on. After completing his PhD on specific aspects of the bioenergetics of mitochondria in an effort to improve organ transplantations, he found himself uncertain about what to do next. So far, so normal. But then things took an interesting turn.

> "I had wanted to do a postdoc, but I had entered a writing competition while I was doing my PhD and I happened to be a runner-up; so, I thought, Well, maybe I can write, then.

> "But I had no idea how to go about writing. So, I looked in the back of **New Scientist** for jobs—either a postdoc, ideally something to do with bioenergetics and mitochondrial function but nothing to do with transplantation, or a writing job—and I happened to get a writing job before I got a postdoc.

> "It was for a small, medical education agency—I had no idea such things existed—and it was basically soft marketing for the pharmaceutical industry, but it was good fun. And it was, in retrospect, very worthwhile: I learned to write that way and I learned to communicate with very different audiences.

> "I realized that you could become an 'expert' in areas you knew very little about, relatively quickly. Now, for me there was a key common factor: for so many of these diseases that I couldn't spell one day, it turned out that when you read about them you'd discover that free radical biochemistry was central to the disease process, so I found myself quickly on home ground.

> "This was really the motivation for writing books down the line: that all of these different diseases seemed to have the same basis in chemistry as to why they were going wrong. So that was thrilling, actually."

Eventually, however, the desire to focus on this exciting insight naturally clashed with the requirements of the rather more prosaic requirements of the corporate world.

"After a couple of years it was really enough, because you had to hop whenever the client said, 'Hop'—there were always subjects that you didn't choose yourself and you never had time to follow through in the kind of depth that you might want to. So I became almost desperate to build on that, somehow—and the ideal way out for me was to write a book that tried to address the deeper, intellectual question of why this free radical biochemistry is underpinning all of these diseases: examining what was really going on there."

That first book, *Oxygen: The Molecule That Made the World*, successfully launched Nick as a writer, so much so that for a time it looked like he was going to forever leave science itself.

*"I had an honorary position at UCL, so I still had a connection with the lab, which was a very useful address for me to have. I published occasional papers and was writing feature articles for **Nature**, **New Scientist** and so forth. I'd written another book, **Power, Sex, Suicide: Mitochondria and the Meaning of Life**, and that was pretty successful in a small kind of a way, I suppose; but I was particularly keen to be addressing these big questions.*

*"I had really developed a lot in how I was thinking over that period of 5–6 years or so, and it was clear to me that there was a limit to how much one can write for **Nature** or **New Scientist** about the same theme: you can do it once every couple of years, perhaps, but they don't want to have an article from the same writer on the same subject every three or four months.*

"So I was faced with a choice: either I was going to become a journalist and write about other people's work and other people's interests or I had to get back into research and really follow through with the kinds of questions that I was becoming more and more interested in. I tried to engage other researchers to do some experiments and to begin to test some of the ideas I had, but I realized that there was no way of making sustainable progress there—people, obviously and reasonably, have their own interests and their own drives."

Finding himself on the outside looking in, increasingly viewed as "a science writer" by his colleagues rather than an active member of the establishment, the opportunity to cross the boundary, as it were, back to frontline research seemed increasingly remote. And then, suddenly, a sort of miracle: the powers that be at UCL decided that a conscious effort was required in order to trigger risky, potentially game-changing ideas that otherwise wouldn't be considered in the naturally conservative environment of standard academe, developing a specific mechanism, the UCL Provost's Venture Research Prize, to shake things up for innovative, unorthodox thinkers.

In 2009, Nick won. And suddenly, he was back in the club, actively pursuing his agenda of demonstrating the overarching importance of looking at an enormous range of key biological questions—from the origin of life to diseases—in terms of fundamental energy-transfer mechanisms.

But through it all, he kept writing: *Power, Sex, Suicide: Mitochondria and the Meaning of Life* was followed by *Life Ascending: The Ten Great Inventions of Evolution*, which was followed by *The Vital Question: Why Is Life the Way It Is?* His next book, *Transformer*, is due to be released in 2021.

Why does he keep going? Well, partly, I suppose, because at this point being a writer is now an essential aspect of who he is. But partly too, it is because the very act of writing clearly helps him sort out his research ideas.

"One of the great things about having written books and thought about this broader picture, is that I now have what I hope is a very solidly grounded picture of how these things fit together across a large scale; and this has led me to conclude that the answer has to be in a certain well-defined area. Maybe I'm wrong in the details, but I'm convinced that it has to be there somewhere. I find that that's true of a lot of these types of questions: people who are in a particular field and are not taking that broader, synoptic view will tend to compare different ideas solely on their own perceived merits or demerits: 'This idea has these strengths and it has those weaknesses, and that one has these strengths and those weaknesses'.

"But the reason that I think the answer lies there has got nothing to do with the quality of the science in the other fields, it's got everything to do with the philosophy that's associated with the very notion of life itself, leading me to ask, 'Well, these proton gradients across barriers are so fundamental to life that they must have arisen early; and they must have happened in an inorganic context originally, so what could that inorganic context possibly be doing?'—a question to which there's only a very limited number of answers.

*"So whether or not it works for **me** in the lab, I'm confident that it **will** work—that having a broader view can funnel you down into a particular question; and even if the evidence for it at the moment is weak, which it is, I think 10 years down the line, the evidence for it will be strong. And if it's not, then I'll be out of a job."*

Well, maybe one job, anyway. But quite likely not that one either.

The Conversation

I. A Long and Winding Road

Nick goes round the houses

HB: I'd like to start with your early years and how you got into science. Were you always interested in biology or even science in general?

NL: I was always interested in biology and chemistry, but I'm rather upset that I was put off physics at school.

HB: When did that happen?

NL: At a very early age—about 13 or 14.

HB: Was it a specific teacher?

NL: It was partly the teacher and it was partly the manner in which the subject was taught; and the kind of physics that I'm really interested in now we never even touched on at school. As a result, I didn't do physics at A level, which I now really regret, because it would have been—well, it's one of my main interests right now, but I never had a formal grounding in it.

At any rate, biology and chemistry, from the very beginning, were subjects that I really loved and I found I was quite good at. It's strange, because I was quite good at English and writing as well, but I never did nearly as well in the exams; and that was, in part I think, because I never knew what they were looking for as an answer.

Science is very straightforward in that sense: the answer is plain—at least it is in exams, anyway. But meanwhile, I think that writing books has helped me to find my way of addressing problems in science, realizing that the exam-type of answers in science—where you're either right or wrong—is actually not really true at all; and it took me a long time to really understand that, which was as much through writing books later on in life.

HB: Was science encouraged in your family?

NL: Well, my father is a historian and my mom was a primary school teacher. My mom had a deep interest in science but never really studied it in a serious way, and my dad is very intellectual and probing in the way that he addresses questions.

Actually, I've often found that, evolutionary biology is not a million miles away from history in terms of how it structures thinking about a question. The actual approach to gathering data or formulating hypotheses is very different, but there's a lot of interpretation that goes on in evolutionary biology which is essentially historical, and I think I get a lot of that from my dad.

HB: He must have been quite pleased by watching your career and trajectory as you moved forwards.

NL: Well, my trajectory has been rather peculiar; it's worked out very well—so far anyway—but it's not a normal trajectory. I was outside science, technically, for some years, writing books. I would love to be able to recommend that to people, because it certainly expanded the way that I see science and I think it's, in fact, very important, to have a very broad view which you can focus down onto particular areas, which is not the way that most people operate.

If you go through a standard degree and focus on a PhD and so on, you become narrower and narrower, but I followed a path that I don't really feel I had much to do with: I just did what seemed the best thing to do at the time; and I've been all around the houses.

HB: Let's talk about that in a little bit more detail. You did an undergraduate degree in biochemistry.

NL: Yes. Which I really did not enjoy. At school, I was massively keen on biochemistry and evolutionary biology and desperate to go to university and study it, but I found it a bitter disappointment. It was probably partly me, it was probably partly that I also left home and found a freedom that was important to me: I spent a lot of time climbing and hitchhiking around and telling stories to people. That was good fun—and, in retrospect, important.

I also found the course, and perhaps the way that undergraduates are taught generally in universities, very constrictive. It was a lot of memory work, a lot of rote, a lot of exams. And now, looking back and thinking how *I* teach undergraduates and how one ***should*** teach undergraduates, all of that memory rote learning has got nothing to do with how science works, how research works or whether or not you'll actually be a good scientist.

It put me off; and it wasn't until I started doing a PhD—and that was, again, a series of lucky accidents, really—I would have left science if I had anything else to do at that point.

But when I started doing a PhD I realized that research was just wonderful; and, since then, I've been going in a more conventional direction.

HB: It's interesting, because my experience partly reinforces yours and partly doesn't, insofar as high school biology was looked upon—by myself and others I hung around with at the time—as deadly boring and nothing more than rote memorization.

My thinking at the time—and this turns out to have been a remarkably common view that I discovered after having these sorts of discussions with other physicists—was that because of that, biology was not only boring, but also considerably *harder* than physics, which was a particularly formidable combination for the lazy teenager that I was.

The sense was that in biology you were forced to memorize all these arbitrary things whereas, in physics you only had to know a couple of basic principles and you could derive everything and didn't really need to memorize anything at all, so it was ever so much easier.

And then there was this aesthetically-pleasing notion of understanding things as a consequence of first principles, which you seem to be very much in favour of as well.

NL: Yes, that's how I come at all these problems in biology: it's very much from first principles, which I suppose is something of a physicist's approach. Actually, a lot of my PhD students are physicists, so there's obviously something going on there.

HB: That sounds quite reasonable to me. I'd like to get back to your career trajectory but, before I do, a specific question: you mentioned the difficulties associated with the undergraduate experience, both from a student and teacher perspective, but did you have a particularly influential high-school biology or chemistry teacher?

NL: Yes, I had several, very good teachers in chemistry and biology, and other subjects beyond that, like history, English and maths. But the two who really stand out most in my memory are the biology and chemistry teachers.

I very often regret that I haven't gone back often enough to say thank you, because teachers set out your life for you in a way that they probably never appreciate—well, they may know that they're doing that, but I doubt that they ever know just how much they influence particular people's lives.

HB: What was it about them that made them special? What sorts of things did they do? How did they engage with the students in a way that was helpful or captivated you?

NL: You would hang onto every word of the chemistry teacher, and it was partly that he put things in an extremely interesting way. It was also partly that, if you were not really focusing, he had an extraordinary shot with the board duster: it would explode right in front of you and you would get chalk in your face. So, you really did focus on his lessons, otherwise you were getting an explosion of chalk in your face—I suppose it was a mixture of the proverbial carrot and the stick, but it meant that you really focused in his class.

HB: It was efficacious, anyway.

NL: Yes.

HB: OK, so let's get back to your story. After you finished your under-graduate degree you weren't terribly inspired by the experience, as you were saying, so what happened next?

NL: I needed to pay off my student debt, and I got a job in a lab that came with the possibility of doing a PhD if I wanted to. I didn't intend to do that, but when I realized how much fun research was, then I really did want to do it. So, looking back, I'm enormously grateful for that opportunity.

The guy who gave me that opportunity is quite a maverick, a guy named Colin Green—my PhD was funded by his winning on the horses for a short period. It would be very difficult to do that sort of thing today; it's become far more formalized.

Talking to various academics at UCL and elsewhere, almost everybody had a slightly unusual ride in one way or another, differ-ent to what we expect our own PhD students to do. I think that we're beginning to lose that freedom to be a bit different, to do things in a different way.

HB: A bit of rebelliousness, almost.

NL: Yes, exactly. It's a shame if that goes, because I think it's an important part of being a good scientist.

HB: Right. Let me ask you specifically about your writing. You've written many books—you clearly have a passion and a gift for writ-ing—and earlier you mentioned the benefits of writing, enabling you to take a broader perspective on things. Did this interest in writing come naturally, as it were, or was it more as a result of what you were reading, you think?

NL: When I was at school, I read books like Richard Dawkins' *The Selfish Gene* and James Watson's *The Double Helix: A Personal Account of the Discovery of the Structure of DNA*. I read popular science books avidly at that age and dreamed, as most kids do, of Nobel Prizes and doing something really valuable and important—and that all kind of drained out of me at university.

Then later, doing the PhD, well, that was different. It concerned a very specific question related to organ transplantation—the bioenergetic aspects of it—how the energy works in the mitochondria, which are the powerhouses of cells. If you take an organ out and store it on ice for two or three days and then transplant it back into someone, there are two problems which arise.

One of them is that you have a potential for rejection and the other one is that, as soon as the oxygen flows back in with the blood again, then the organ is likely to go wrong: it's called ischemia reperfusion injury and that was what my PhD was about.

So, it was strongly geared towards solving a problem, a very specific problem to do with kidney transplants, and one that would make quite a difference in the world if you could do it because we can only store these organs for two days, at the most, in the case of kidneys; for a liver or a lung it's a matter of hours.

So I was quite driven to find a solution, and the reason I ended up leaving science again was that I never did come up with anything like a solution. I had a great time doing the research, but I didn't get anywhere close to where I wanted to be with it, and I needed to leave that field because it was plain that I had become stale—I didn't have an alternative view of it.

HB: I see. And then from there you became a professional writer. How did that happen?

NL: Well, I had wanted to do a postdoc, but I had entered a writing competition while I was doing my PhD and I happened to be a runner-up; so, I thought, *Well, maybe I can write, then.*

But I had no idea how to go about writing. So, I looked in the back of *New Scientist* for jobs—either a postdoc, ideally something to do with bioenergetics and mitochondrial function but nothing to do with transplantation, or a writing job—and I happened to get a writing job before I got a postdoc.

It was for a small, medical education agency—I had no idea such things existed—and it was basically soft marketing for the pharmaceutical industry, but it was good fun. And it was, in retrospect, very worthwhile: I learned to write that way and I learned to communicate with very different audiences.

A lot of it was doing animation of the mechanism of action of drugs, so you would have to set things up by saying something like, *"Here's what's going wrong with neurons in the brain as one degenerates in Alzheimer's disease, and here's how this drug interacts with that picture"*.

You would have a 10–15 minute, 3D animation of this process. I would write the narration and envision the camera angles, and then work with an artist and a producer and animator to make these things. It was great fun. It would be different jobs over different weeks; I would put in proposals for new work.

HB: It was probably quite a learning experience too, I imagine, in many different ways. There's learning about how to write effectively and reach your audience, but if you're bouncing around from topic to topic—doing Alzheimer's one day and something else the next—you must have had to read up on quite a wide variety of different things.

NL: I realized that you could become an "expert" in areas you knew very little about, relatively quickly. Now, for me there was a key common factor: for so many of these diseases that I couldn't spell one day, it turned out that when you read about them you'd discover that free radical biochemistry was central to the disease process, so I found myself quickly on home ground.

This was really the motivation for writing books down the line: that all of these different diseases seemed to have the same basis in chemistry as to why they were going wrong. So that was thrilling, actually.

I also had to learn to write for an international audience where English was generally the second language; and that means cutting out any flowery, verbose English and writing not only plainly and clearly, but also in a continuous forward-moving direction.

Because you can't set it out as you would in a scientific paper saying, "*Well, it could be for this reason or that reason*". You can't have a discussion that flows sideways, if you like; instead you have to say, "*This leads to this, leads to this, leads to this*", because the camera is tracking that.

That forces you to tell a straight story; and that was also an interesting learning experience. So, I learned a great deal from that experience. But after a couple of years it was really enough, because you had to hop whenever the client said, "*Hop*"—there were always subjects that you didn't choose yourself and you never had time to follow through in the kind of depth that you might want to.

So I became almost desperate to build on that, somehow—and the ideal way out for me was to write a book that tried to address the deeper, intellectual question of why this free radical biochemistry is underpinning all of these diseases: examining what was really going on there.

HB: That's interesting, because I had naively assumed that your present "bioenergetic focus", as it were, was the natural result of your exposure to mitochondria as a PhD student, reflecting on the broad-based impact of this wonderful ability to convert energy in a very efficient way.

NL: Yes: it took me a lot longer. I had my first intimations, when I was working with this medical communications agency, that free radicals really were important; and that the kind of chemistry and biology that I knew about underpinned an awful lot and was worth trying to explore further.

But I didn't really know why or how; and it wasn't until I started writing my first book, *Oxygen: The Molecule That Made the World*, that any of this really came to me. It was the process of writing the book that gave me—I'm not really sure I can say "deep" insights—but a far deeper, broader-based understanding of the world than I ever had before.

HB: Describe a little bit of the sociology of the field to me. I say this as a complete outsider; but my sense is that when you take a "big picture" view of biology, a large chunk of that these days consists of people looking at things from the perspective of genomics and how the vast amount of resulting information associated with that can be recognized and understood—both independently, as it were, but also in terms of how the resulting protein processes and so forth interact with the surrounding environment.

And here you come along saying something like, "Well, that's very important, of course—indeed, essential—but it's not the whole story; we also have to look at things more from a bioenergetic perspective, because that imposes essential constraints on these other factors and it's deeply tied to them."

Were there—are there?—other people thinking along those lines, or does this put you somewhere near the fringes?

NL: There are some people thinking along these lines, yes, but not many. Probably the person who's influenced me most—I first met him some years later, after I'd written that first book on oxygen—was a guy named Bill Martin.

Now, he is really a geneticist by background but, unlike most geneticists, he's extremely interested in biochemistry and physiology and what genes really do in the context of what they're doing it in; and he's surprisingly unusual in that.

He has enormously sweeping and inspiring ideas about the evolution of life. The first time I met him I thought he was completely mad. I didn't buy his ideas at all; and now, I do. It took some time to come around to them because they were so radically different.

I met him outside a conference hall; he was about to give his talk in a few minutes and he was having a cigarette outside. He's a big, brash Texan guy, and I remember that he was trembling a little bit—he was obviously very nervous—which seemed surprising to me given that he didn't look like the nervous type.

And then he got up and gave this extraordinary talk, in which he argued that the Bacteria and the Archaea emerged separately from hydrothermal vents and that those first cells didn't have cell membranes as we know them—the whole thing was almost impossible to take on board, it was just so utterly different to anything I had ever heard before; and by then I had read widely about early, evolutionary history.

HB: What was the reaction by others at the talk?

NL: Stunned silence, really. I mean, he's quite an abrasive character and he has a knack of making enemies even of his friends sometimes; and that's a real shame because he's a wonderful man.

In short: these ideas were right on the outside of a normal place to be, but they're very solidly grounded in what we do know about biochemistry and physiology and genetics—so I think he's right.

It took me a while to see the world from closer to his point of view—and it's not as if we share *all* our thoughts on it even now—but he's probably the one person who has taken a similarly broad bioenergetic view of the whole sweep of evolution.

HB: Very good. I'm almost ready to go into those details, but before I do let's get back to your story to bring things up to date. I left you, as it were, after you wrote your book, *Oxygen: The Molecule That Made the World*. What happened after that?

NL: Well, throughout this whole period I had an honorary position at UCL, so I still had a connection with the lab, which was a very useful address for me to have. I published occasional papers and was writing feature articles for *Nature*, *New Scientist* and so forth.

I'd written another book, *Power, Sex, Suicide: Mitochondria and the Meaning of Life*, and that was pretty successful in a small kind of a way, I suppose; but I was particularly keen to be addressing these big questions.

I had really developed a lot in how I was thinking over that period of 5–6 years or so, and it was clear to me that there was a limit to how much one can write for *Nature* or *New Scientist* about the same theme: you can do it once every couple of years, perhaps, but they don't want to have an article from the same writer on the same subject every three or four months.

So I was faced with a choice: either I was going to become a journalist and write about other people's work and other people's interests or I had to get back into research and really follow through with the kinds of questions that I was becoming more and more interested in.

I tried to engage other researchers to do some experiments and to begin to test some of the ideas I had, but I realized that there was no way of making sustainable progress there—people, obviously and reasonably, have their own interests and their own drives.

HB: So, you found yourself on the outside looking in, and you had to find a way to get back inside.

NL: Yes. And at that time—this was 2008—UCL announced a very unusual prize that was called Provost's Venture Research Prize that was available for any researcher within UCL.

It was couched in very grand terms—*Potentially transformative ideas that change the way we think about an important idea; Nobel Prize-winning-type ideas*—but I think most people saw it simply as potentially mad ideas that are unlikely to get funded by the research councils.

And if you see it in *those* terms it changes your perspective. So rather than thinking, *I don't have any ideas on that sort of grand scale*, you instead say to yourself, "*Well, mad ideas that are unlikely to be funded—I have a whole book full of those*", it helps.

I figured I had nothing to lose and everything to gain from this, so I put in a short proposal—it was literally one page, which was all that was asked for—and it bounced back to me from the guy who was running it, Don Braben.

Don was of the opinion that the peer review process for proposals was killing really innovative, breakthrough-type research, because the people who review it, being your peers, have invested their whole careers in a particular way of seeing the world.

He's thinking about major, major breakthroughs—people of the calibre of Einstein and Planck and so on. He believes that there have been several hundred of these over the 20th century, people who just see the world in a different way but who, now, would be killed by the modern approach of, *You've got to get the backing of your peers to get X amount of funding*—they simply would not have survived in that kind of environment.

I think he's probably right, actually; and this was his attempt within UCL to try to back that kind of research.

HB: So, you wrote this one-page proposal.

NL: Yes. And then he came back at me with all sorts of criticisms and comments, saying things like, "*Address these, and we can talk again*". It was a very iterative process: I met him on three or four occasions, and each time, he batted me back with a whole bunch of problems and questions. Eventually the proposal swelled to about 20 pages, by which time he said to me, "*Okay, I think you've answered all of my questions and I'm willing to back you, but nobody's going to read this proposal: get it back down to four pages and someone might read it*".

Then it went to David Price, the director of research across UCL. I had an interview with him and then with the Provost himself. At no point was there a conventional peer review process, at no point was I interviewed by biologists who had any kind of vested interested in this.

It's not, in fact, a million miles away from how universities appoint staff: people give seminars and then have an interview with a panel, which is not made up of experts in that particular subdiscipline, but clever, experienced types who, between them, can more or less cover a range of ways into that person's subject.

And from that one can usually judge reasonably well, I think, whether or not someone is any good or whether the ideas are coherent. So, I don't think this approach is actually very far away from the mainstream in many ways; and it shouldn't be.

HB: Well, kudos to UCL for doing this, because in my experience so many institutions become, to some extent, the victims of their own success: once they develop a large reputation, the price all too often paid for that reputation is a sense of conservatism, an increasing amount of unwillingness to ensure that anyone outside of the mainstream who risks sullying that reputation is kept outside.

That UCL is exhibiting an explicit willingness to shake things up and thereby countenancing the possibility of involving people who are not necessarily already within the establishment or the mainstream takes a good deal of courage, foresight and independent thinking. Very good for them.

NL: Absolutely, I hugely benefited from it. And you're right: this scheme is not running in any other university in the UK.

HB: Do they still do it?

NL: Yes, absolutely: it's a unique place. The fellow who's really running it, Don Braben, is always willing to meet people. He tends to bat them away; and if they keep coming back, he keeps batting them away until he becomes convinced that the person has an idea that's really worth something.

HB: OK, but that strikes me as an important part of the process too—not only to sufficiently convince him, but to ensure that the person is sufficiently tenacious and determined, while being flexible enough to listen to what a knowledgeable and objective person has to say about their ideas.

NL: Yes, that's right. And the ideas have to be open-ended; it's not like testing a particular hypothesis that could have a yes or no answer at the end of it, it's really about a particular way of seeing the world that could change the world.

I mean, to put it in its simplest terms: energy and enthusiasm matters too. And that can't possibly be wrong, can it?

Questions for Discussion:

*1. In what ways does the peer review process effectively **decrease** innovation? Readers interested in a more general assessment of the history of peer review are referred to Chapter 6 of **Science and Pseudoscience** with Princeton University historian of science Michael Gordin.*

2. Should young scientists be encouraged to have other life experiences rather than moving rapidly through their degrees?

3. Is "rebelliousness" a desirable quality in a scientist?

II. Structuring Energy

Cells, membranes and a counterintuitive mechanism

HB: So this seems an appropriate moment to turn our attention to *The Vital Question* and exploring that big idea you were talking about in more detail.

Let me start by giving you a brief précis of what I took away from the book as a complete non-specialist and you can tell me if I'm right or wrong before you go into various aspects of the details.

The book captivatingly engages right from the beginning by focusing on—somewhat unusually for popular science books—open questions: that is, what we *don't* know rather than what we do.

So here is my sense of some of these open questions and the way—or at least one way—you suggest we go about addressing them.

We are in this situation where we have three fundamental types of cells, and every living being around us is made up of one of these three types. Two of these three are called "prokaryotes"—Bacteria and Archaea—and one open question is understanding how, exactly, they are related to each other. Meanwhile, there's the question of how the third type, a more complex type of cell called a "eukaryotic" cell comes into being in the first place. The standard story seems to be that these prokaryotes were around for a few billion years doing their thing and then, suddenly, these eukaryotic guys show up, and it's quite unclear how that happened, exactly.

Then there's a more general question related to the energy process in all of these types of cells. It seems that they are all effectively powered by something called proton gradients—protons across membranes—and there are a bunch of related questions associated with that: Why is that? Why is this so remarkably universal? What is this telling us about the underlying processes that give rise to what we call "life"?

In particular, one related aspect to focus on is the question of stability, both in terms of this energetic process itself and the life-forms associated with it. The very fact that we have bacteria still around four billion years later seems quite telling.

In short, the idea seems to be that focusing on the specifics of this energy mechanism, these proton gradients, will allow us to more concretely address and hopefully understand these larger issues.

That's my sense of things. That's what I picked up.

NL: That's a very good synopsis.

HB: Okay, good, so I haven't completely struck out. I should add parenthetically that this is undeniably a great book that I would wholeheartedly recommend to anyone. Go buy it.

OK, end of advertisement. Now it's over to you to sketch out the main arguments before I ask you more questions about things.

NL: Sure. So the book is called, as you mentioned, *The Vital Question*, and the subtitle is, *Why is Life the Way It Is?* That's really what I have in my mind as "the vital question", but it's not obvious from the outset what that question actually means.

What it means is that all complex life that we see around us—plants, animals, fungi, basically everything that you can see—is composed of a very particular cell type; they're large, complex cells with all kinds of bits and pieces inside that they share.

If you look at one of our own cells down a microscope, you would struggle to tell the difference between that and the cell of a mush-room or a plant outside the window: they're remarkably similar. You could list page after page of the detailed traits that they have in common.

And that's very, very different to the structure of either Bacteria or Archaea—this other type of prokaryote. Archaea are small, simple cells that, like Bacteria, don't have a nucleus. They look like Bacteria but, in fact, in their genes and their chemistry, they're actually quite different.

HB: You mentioned that they're named wrongly too, because you'd naturally assume that they were the oldest things around.

NL: You would think so; and that's why they were named that way—but they're almost certainly not any older than Bacteria; they're probably about the same age.

Life on earth has had a very peculiar trajectory: it started very early, so far as we can tell—about 4 billion years ago—and then we had, as you mentioned, this 2 billion year delay before anything really interesting happened—interesting at the level of large, morphological complexity.

Then, this one cell type arises—once, it seems to have happened once in 4 billion years—and we have this enormous explosive radiation of the major different groups of eukaryotic cell types: not only the plants, the animals, and the fungi, but also thousands of single-celled things like amoeba—that's where the real variation within the eukaryotic world is—but they're all basically the same cell structure.

The things that we have in common are things like sex, for example—having two sexes—and going around and engulfing other cells and eating things and so forth. These are all traits that all these cell types seem to share, and it's not obvious why none of them arose in Bacteria or Archaea. And it's also not obvious if it would necessarily be this way on another planet.

So here I'm approaching the physicist's view again, as you mentioned earlier: trying to predict from first principles what we might expect to see as life on another planet.

Some of it is straightforward—*Would it be carbon-based and need oxygen?*—but some of it is quite strange: *Would it be sexual?*

We don't know. It's a 100-year-old problem of why life is sexual on earth; and the strange thing is that entire field, for 100 years, compared sexual organisms with asexual organisms—things that clone themselves, like dandelions and other plants and animals.

It turns out that if you clone yourself you can grow much faster and do much better for a period of time but then they all fall extinct. And it's never been entirely clear *why* they all fall extinct, but it seems that there's just a lot less variation in them, so they're far more likely to get wiped out by a virus or something.

HB: Right. But let's return to this "first principles" perspective you were talking about a moment ago. It seems to me that the obvious question is this: despite the fact that when we look around us we see such enormous diversity in species and subspecies and all that, nonetheless all these wildly different life forms all seem to be composed of essentially these same type of these eukaryotic cells—so why is that? What's going on?

NL: That's exactly it, that's this vital question. Now, what's going on?

In a very short answer, what *I* think is going on is that Bacteria are simply structurally constrained: they're very small, simple cells and they breathe, in effect, across their membrane. They breathe through their skin while Eukaryas don't—we've internalized it in mitochondria, which were Bacteria once.

So, some prokaryotic cell swallowed another prokaryotic cell, and the long-term effect of that was that it was breathing internally— it had internalized the whole respiration process. But it's not very obvious why that would make such a big difference, or why bacteria can't just do that sort of thing without first swallowing a cell.

Now, the swallowing the cell, I think, is the critical factor—it's not so much about one cell, actually. There's a lovely quote from the French biologist Jacques Monod, who said that, "*Every cell dreams of becoming two cells*".

Cells just become populations—most people have seen populations of bacteria dividing away madly. So, you have a cell within a cell, and it very quickly becomes a population of cells within a cell; and they compete with each other to get into the daughter cell—into the next cell—and so on.

What you have then is a population of living things inside a population of cells; and that is probably the single-defining moment of the whole history of life—which is that, in very simple terms, it changes the selection pressures.

So, what you have to do to survive is not worrying about the outside world, it's about worrying about the "inside world". It's for those reasons that *all* these eukaryotic cells have got *all* the same traits, because they were *all* dealing with this same problem very early on in their evolution: How do I deal with the population of bacteria that's going to tear me apart if I'm not careful? How do I constrain it? How do I keep it doing useful things for *me*?

HB: Right; and perhaps I'm mixing metaphors a bit, but to me this whole approach of distinguishing between "inside" and "outside" necessarily brings up this whole notion of a membrane: right from the beginning, as it were, you are looking at things from a "membrane-centric perspective" as it were.

NL: That goes right back, again, to the origin of life, because Bacteria, as I say, are respiring across their outer membrane—we've simply internalized all of that, along with the genes needed.

There's a lovely quote from what you might call the "founding father" of bioenergetics as a field, a guy named Peter Mitchell who won the Nobel Prize in 1978, and he had a paper on the origin of life in 1957 at a conference in Moscow.

I probably won't be able to remember the quote verbatim, but he was essentially saying that he sees life as being the dynamism of things crossing the membrane, of phases on opposite sides of the membrane; and you can't really tell the difference between what's the "inside" and what's the "outside"—they're simply two phases on opposite sides of a structure which is both separating and uniting them.

When you begin to see the world in that kind of way, and you think about the structure of hydrothermal vents and you think about the structure of energy flow in living cells, you realize that it's an extremely productive way of seeing the world.

HB: Okay, so I'd like to discuss these hydrothermal vents and all that in more detail later, but first I'd like to focus on the act of cellular respiration and this "proton-motive" force.

As I understand it, all of these three different types of cells we were discussing earlier undergo some form of this respiration. And from a chemical perspective, what it amounts to and what it, basically, amounts to is, from a chemical perspective, stripping electrons from food.

NL: Yes. We basically have a current of electrons in the membrane stripped from food and passed to oxygen, so we literally have a current.

HB: And that current that's created is what is responsible for pushing protons outside of the membrane.

NL: Yes: that current is driving the extrusion of protons across the membrane. And then we have something like a hydroelectric dam, where the membrane is the dam, and we've got a reservoir of protons on one side of it.

And then there's the turbine in that dam, which is the ATP synthase, the energy-generating protein. It makes ATP, which is often called the "energy currency" of life, but you could really think of it as just cash to spend on things—"energy cash".

HB: OK, fine. So I'm coming to a question—just bear with me. Part of my problem is that I don't understand any of the details, which is a decidedly less grandiose way of expressing things than saying that I'm a "big picture" guy.

But anyway, this is what I'm thinking. You mention in your book that this current can be evaluated in terms of quantum-tunnelling and associated probabilities, which certainly seems reasonable to me.

And that makes me think something like "*Well then, wouldn't it be the case that the strength of this current—and the associated effects, like how well it pushes protons across the membrane—depend on some structural aspects of the elements or molecules or whatever that it's tunnelling through?"*

I mean, I can imagine creating some sort of mathematical model, say, when you can say something like, *"If we space these things at this level, they wouldn't be as efficient, if we use this other type of material, it would be less efficient or more efficient"*, or something like that.

I guess what I'm struggling with is the possible variation of efficiency within what seems to be a strikingly universal process.

NL: Well, electrons do hop, by quantum-tunnelling, from one centre to the next, and the likelihood of them doing that depends on the distance, so it has to be a small distance—in the order of 12 angstroms or so.

So, it depends on that distance. It depends, as well, on the pull from there—the reduction potential—the degree to which it wants that electron; and it depends, as well, on whether it's "got room for it, chemically", in effect.

These factors are all under selection—you can see that the distances between the centres are really the same between wildly different species.

They're set in very large proteins and the proteins have these three-dimensional structures; and within that structure, it has this quantum-tunnelling site here, and then maybe there's another protein sub-unit over there somewhere that holds the next one back over there, say—and what's been selected for is that distance between them is 12 angstroms.

And then there's the thing which is pulling it through and making it hop from one site to the next is the "sucking power," if you like, of oxygen, which is down at the other end. It doesn't have to be oxygen, it can be almost anything—iron, or nitrate, anything which wants to grab an electron will work—it's just that oxygen does it particularly well.

HB: Right. So let's get back to what you were saying earlier when you mentioned the possibility of extraterrestrial life, examining things from a first-principles approach where you're effectively trying to distinguish *Does it **have** to be this way? Is this part of some type of necessary, guiding principle for life?* from *Did it just **happen** to be this way? Is this a sort of accident—or at least one possibility out of many?*

So, I'm thinking, *Why this particular mechanism?* It seems particularly important to note that *all* cells are doing this sort of thing, to some extent—that seems highly suggestive of some kind of universal law.

NL: Well, the question is, *What **does** it suggest?* I don't think we know, and this is not the kind of question that many people worry themselves about.

It could suggest one of two things: it could suggest that it's just really good, and so it spreads and it's spread across the whole world because it's better than any other way of doing it.

Or it could suggest that there's a fundamental reason underpinning it that it *had* to be this way and other ways either don't work as well or simply never got a hold in the first place. In fact, it's quite possible to think of alternative ways of structuring energy.

HB: Okay, good: that's what I wanted to get to.

NL: It really does look as if there's a peculiarity about this: it's one of the most counterintuitive ideas in biology, one that Leslie Orgel, a famous chemist working on the origin of life, called, *"The most counterintuitive idea in biology since Darwin"*.

It's different to something like DNA, where as soon as you see the structure, you understand how it works. And with this, it's just not chemistry.

Everybody thought in terms of chemistry—*you want to activate this molecule to make it react, so you transfer a phosphate group onto it and that makes it a bit more reactive and then it does its thing.*

But nobody even dreamed that the intermediate would not be just another chemical molecule, but would be a membrane with an ionic gradient across it.

It's odd; and I think, in a way, that's the joy of the whole thing, because, *Why is it like that?* Is it just an oddity or "quirk" of the particular origin of life here? Is it something which happened a bit later and spread for some weird reason?

What I'd like to think, and what I think **is** the case, is that it's fundamental: that not only does it structure the way that life works here—from the origin of life, right the way through this long period of stasis of bacteria, through to this abrupt, almost unique origin of all complex cells, through to all of the complex properties of these cells—but that it alludes to some general principle that can in some ways explain that whole continuum.

It almost certainly doesn't—nothing in science works like that—but it would be wonderful if it could restructure the way we think about these questions; and I think it might.

Questions for Discussion:

*1. To what extent do "historical sciences" such as evolutionary biology, geology and cosmology fundamentally involve different approaches than other areas of science? Readers interested in this topic are referred to Chapters 1–2 of **Astrophysical Wonders** with Scott Tremaine and Chapter 8 of **Science and Pseudoscience** with Michael Gordin.*

2. Might our sense of the "counterintuitive" nature of the energy membrane mechanism described in this chapter be related to how we classify phenomena into divisions like "chemistry", "physics" and "biology" in the first place?

III. Hydrothermal Vents

More than just chemistry

HB: I'd like to take you back to where I cut you off a moment ago, when you were talking about hydrothermal vents. My understanding is that now we're talking about the origin of life and by "origin of life," we're looking at the origin of both these bacterial and archaea cells, because as you were saying the eukaryotic cells came a couple billion years later, so we don't have to worry about them for the time being.

Perhaps I can ask you to begin by describing these early experiments in the origin of life back in the '50s with lightening and a flask of so-called "primordial soup".

NL: Well, there's been an intellectual tradition in the origin of life of synthetic chemistry and what works, going back to that Miller–Urey experiment in the 1950s.

Even in the 1950s, the idea that the earth was like Jupiter in that it had gases like methane and ammonia and hydrogen wasn't exactly discredited, but most serious geologists at the time thought that the early earth was full of carbon dioxide with those gases never really geologically likely, but chemists don't worry about that kind of thing; they're interested in chemistry that works.

And if you start with CO_2 it doesn't work: you try to get it to react with hydrogen and nothing happens. And that's important to know, because if you *could* make them react, you would be able to strip CO_2 out of the atmosphere, reversing global warming, and perhaps convert it into synthetic gasoline and solve the energy problem as well.

So, if we ***could*** make that work, it would have a massive impact on the world, but unfortunately it doesn't.

Now, it's very difficult to publish stuff that doesn't work. The chemists have done wonderful chemistry, starting at first with things like methane and ammonia. They accept that the earth probably didn't start that way, and so they began thinking about things like cyanide and formamide—relatively reactive molecules—and they have been able to synthesize things like nuclear dyads, the building blocks of DNA, from them.

This gives you an impetus, then, because it works: everything looks as if it's falling into shape—this chemistry will produce the building blocks of life; and, sometimes, only those and at quite high yields. It's hard to gainsay it, because it's good, experimental science; but it's nothing like life—and I think they've been guilty of ignoring biology.

They take biology into consideration insofar as they say, "*Okay, we need to synthesize DNA and RNA and proteins from very simple molecules. We don't know what the early earth was like, we don't know how long a time there was between that chemistry and the first cells, so it's intellectually, perfectly acceptable.*"

All of which is true. But what they've come up with is a series of pathways that just doesn't look anything like a cell as I recognize a cell. They haven't really closed the distance between prebiotic chemistry and a cell, as we know a cell.

Now, the reason I became very interested in this is because of these alkaline, hydrothermal vents, where you have natural, proton gradients across inorganic barriers, so it's very analogous to cells.

It throws up this possibility that cells have a structure—and this is another thing that the chemists have not worried about very much; they tend to do reactions in a test tube in solution and there isn't structure there—but cells have a structure: everything is going on across this membrane. And if you ignore that membrane, then maybe you've ignored the most important thing.

Then, the question is, *If we're doing prebiotic chemistry, we don't **have** a membrane, so what do we do?* Well, what *do* we have? We have an inorganic barrier and fluids on opposite sides which are very different in their chemistry and will react if they get in touch with each other.

HB: So, we have a large-scale analogy, something which looks very much what is required at the cellular level.

NL: Yes, and we have a continuous flow as well going through it.

The difference between living and not being alive is really energy flow: when someone dies, there's really no difference between when they were alive and when they were dead except they no longer have any energy flow—they've not lost any genetic information in that moment. And this is really at the heart of the origin of life: all cells have this continuous energy flow; and they waste a tremendous amount—excretion is massively over-looked.

So there is a continuous, chemical reaction, which is powering everything; and just by continuously breathing oxygen, eating food, we're putting out waste all the time. All cells do that; and this is what vents do as well: they're continually feeding in hydrogen in these alkaline fluids and venting out their waste.

There are various ways in which it's analogous to a cell, but it's purely inorganic, so the question is then, *Well, how might these forces of chemistry and physics be harnessed to driving the growth of protocells, the first cell-type things?*

But at least looking at things in this fashion you now have the possibility that there may be some way in which these proton gradients across inorganic barriers *could have* led to this mechanism being taken over by cells, becoming fundamental to cells, and being used by all cells on this planet.

So there are explanations for why that might work—I won't go into them in detail now—but the key point is that these are *testable*: there are very simple questions that can be addressed in the lab. We've built a reactor to try to test some of them—it's very simple, it doesn't cost much, and it's pretty amateurish, I have to say—but nobody else has ever tried to ask those particular questions about the structure and about whether hydrogen can convert CO_2 into organic molecules if driven by a proton gradient across a barrier. So that's the essence of it.

HB: Wasn't there this guy from JPL... I can't remember his name at the moment.

NL: Mike Russell.

HB: Right. He had suggested this very idea, if memory serves.

NL: Yes.

HB: And incidentally, he seemed to be another one of these cranky, rebellious types. Somehow a good many of the colleagues in your field seem to often have that sort of profile.

NL: It does seem that way, doesn't it?

HB: And he had developed this idea before such hydrothermal vents had actually been found, right?

NL: Yes, exactly. Mike Russell had laid out a lot of the theory underpinning the idea that these alkaline, hydrothermal vents could effectively be an electrochemical reactor for the origin of life. He first started putting that forward in 1989 and had a series of papers through the early '90s where he drilled down into exactly how it might work. All of this was very much on the periphery of the field, and it wasn't until the discovery of Lost City in the year 2000, that suddenly, it became mainstream: he took on, almost prophet-like proportions, as he had predicted exactly what these vents would be like.

HB: Where is this Lost City, exactly?

NL: It's off the mid-Atlantic ridge. The key point is that's it's not *on* the mid-Atlantic ridge, because these are not volcanic reactions that are driving it: it's a reaction between cold rock, in effect, and water, with the ocean water percolating down into the crust.

So, the idea is that it would happen on any wet, rocky planet; and that means billions of exoplanets across the Milky Way.

HB: OK, but that presumably also means that we should have lots more of them here. How do you actually go about finding these things?

NL: Well, this was discovered by accident. We know that the mid-Atlantic Ridge is an extremely interesting place—lots of expeditions go there. It's not easy to get there: you're several kilometres beneath the surface.

As it happens, Lost City was discovered by an expedition that was going to the mid-Atlantic Ridge; and about 20 kilometres away some PhD student—it was her first trip down, I think—looked out of the window of the submersible and said, "*Wow, what's that?*"

Part of the reason that it was discovered so late was that nobody had bothered to look. There's a natural tendency to go to the "interesting places", and you don't really get those kind of vents on the ocean-spreading centres where you've got the magma directly welling up very close to the surface, you tend to get them further away. But you tend to get them in that region precisely because of the spreading centres: the ocean crust is spreading across—this is fresh crust, which has been exposed to the ocean water—and it's coming from regions of the mantel, which are quite close to the crust.

So, it's the type of rock that will react with water: it's rich in magnesium and iron and things like that and it's basically just reactive.

Now, if you didn't have any tectonic spreading, then the entire ocean crust would—the technical term is "serpentinize"—it would react with water and turn into a different mineral: it would metamorphose into a different mineral called serpentinite, and that would happen down to a depth of 5–10 kilometers down and everything would stop, that would be it.

So, the fact that we have plate tectonics is a necessary condition. We have a living planet; and I think it's important to see this continuity between the continuous movement of geology and the continuous movement of living cells—the processes are very similar.

HB: Very interesting. So the idea is that these hydrothermal vents give you something that's similar to this proton gradient that's structurally key for cell biology, enabling us to imagine the development of both Bacteria and Archaea from there. And what separates Bacteria from Archaea, if memory serves, is that they have rather different types of membranes. Is that right?

NL: Yes. There's a paradox there. There are these two major domains of life, the Bacteria and the Archaea. They look the same, they both arose 4 billion years ago, they both have effectively infinite population sizes over virtually infinite periods of time; and yet in their morphology they are very constrained: they remain tiny cells without much morphological complexity with fairly small genomes and so on.

And so it becomes an interesting question: *What constrains them?* We've touched on that already—that effectively they're respiring across their membrane, and that's what provides the constraints and stops them from becoming larger and more sophisticated.

But we have this paradox about the origin and divergence of the Bacteria and the Archaea. We think that the common ancestor was in these hydrothermal vents; and we're fairly sure that all the Bacteria and Archaea are using proton gradients across membranes to drive their growth—so it's not just energy metabolism, it's carbon metabolism as well—but the thing is that they have very different membranes, and so there's a paradox there about what kind of membrane their common ancestor had.

Was it one type that got replaced somewhere in the process, or did it have something which was different?

We've done some theoretical work—which would be appealing to physicists, because it's basically calculating the proton flux through membranes and through the proteins and so on—just to work out if it's feasible that natural proton gradients could drive these processes in living cells, because it's not obvious that they can.

What you would expect is somewhat different. Let's say you've got a membrane and a turbine in that membrane—like the ATP synthase we spoke of earlier—and you've got lots of protons on one side and relatively few on the other, then the protons will come cascading through until you've balanced the concentration or the charge across the membrane and you've reached equilibrium—and then, it will stop.

This is what most bioenergeticists will say: that it will never work because you're always going to get equilibrium in no time at all.

Now, in a vent you've got this continuous flow from this alkaline fluid: so then the question becomes, *What is the rate of flow? Under what conditions could it work and when would it not work?*

The long and short of all of it is that it can work very well as long as the membrane is really leaky to protons.

So, they can come cascading through these proteins, but also through the membrane, but because you've got an effectively infinite ocean of protons out there, when they come in, you've still got more to replace them; and you've got a continuous flow of hydrothermal fluids, which is continually taken them away as well—and so they're flushed out across this very leaky membrane.

So it does work if you've got a very leaky membrane—in other words, a membrane which is not like either the Bacteria or the Archaea but an ancestor which is simpler and necessarily very leaky.

Questions for Discussion:

*1. In what ways do reflections on the origin of life help yield a deeper biological understanding to what life actually **is**, biologically-speaking, and what, exactly, is the dividing line between "inorganic" and "organic"?*

2. What do you think Nick means, exactly, when he talks about "the continuity between continuous movement of geology and the continuous movement of living cells"? Might a deeper understanding of biological processes inform our understanding of geological processes as well?

IV. Simulational Challenges

Making your own hydrothermal vent

HB: I'd like to turn to some of your experiments that you alluded to earlier to simulate the processes in these hydrothermal vents. I understand that in one of your experiments you've actually been able to generate formaldehyde—which doesn't actually mean much to me, I have to admit, but might well be highly significant to someone who actually knows about these things.

NL: Well, we're trying various different avenues—that's one particular set of experiments and that's probably the main thrust. Yes, we have succeeded in making formaldehyde—that's not a big deal in one sense, but it's a big deal in another sense.

It's a big deal because, as I said earlier, hydrogen and CO_2 normally won't react with each other—you can mix hydrogen and CO_2 together in a flask and pressurize the whole thing and put a match to it and nothing is going to happen—but if you *can* get them to react with each other, then they release energy in reacting.

In other words, there's a barrier to their reaction, but once you've overcome that barrier, then they should react quite readily; and the thermodynamics tells you, in effect, that under these conditions, you will make cells.

There will be other barriers as well, but what the cells that live in these vents that grow from this reaction do is to use proton gradients across the membrane to drive that reaction.

And the way they do it is to use iron-sulphur proteins. Many proteins have an inorganic group right at the heart of it, it's that that's really driving the reaction; so the protein's providing the context for it and it's drawing in molecules and bringing them in contact in the right orientation with this ion cluster at the heart of it.

These are, effectively, mineral structures that exist in vents. And the expectation is they would have been in those early vents and be able to drive, catalyze, some of these early reactions.

Now, they do to a point, but if you simply mixed iron-sulphur minerals with carbon dioxide and hydrogen, nothing much happens either.

So the question we had was, *"Well, is it the proton gradient that makes the difference? Is it the structure, the barrier, the fact that you've got these iron-sulphur minerals sitting in a barrier with fluids on one side of a particular acidity and on the other side a different acidity and the barrier acting as a semiconductor, transferring electrons from one side to the other side?"*

So that's the setup that we've tried to emulate in the lab. And it turns out that the biggest barrier to the reaction is the first couple of steps to get to formaldehyde. It's not that we're particularly interested in formaldehyde in itself, but we're wondering, *Can we get to the top of that barrier? Can we clear the barrier?*

HB: OK, I understand the concept of finding a way to clear an energy barrier to get a process going. I'm wondering about another issue, though: that of timing and efficiency appropriate to these evolutionary timescales.

I mean, I can imagine someone saying, "Yes, that might work in principle, but you've only got a trace amount of what was needed under the circumstances—you should have had much more. You should have had this percentage or that percentage, given the amount of time that's involved." Is that the sort of calculation or modelling that goes on as well?

NL: No, not for me. Christian de Duve, a Nobel Laureate who worked on the origin of life, made a very important point in this respect, which is that life is about chemistry, and chemistry happens quickly or it doesn't happen at all. If a reaction doesn't happen quickly enough, something is going to take away the reactants, or they're going react with something else: you have to have reacting right here and now or it's not going to happen.

So thinking in terms of millions or billions of years is not actually very helpful. It might take millions or billions of years for evolution to happen, but most of evolution is about stasis, not about change; and change can happen very quickly.

The fact that Bacteria remain Bacteria for 2 billion or 4 billion years doesn't say that they're trying to change, it says that selection keeps them as they are; and we don't necessarily know how long it took for the selection pressure to change and for complex cells to then arise. The assumption is that it takes millions and millions of years but it could be tens of thousands of years—we don't really have a feel for time in biology.

That goes even more for the origin of life.

My feeling is that at least producing organics had to happen quickly and on a large scale; and any system which is doing that is getting you into the right ballpark—and then the difficult parts related to self-organization of matter come into play.

For a membrane that spontaneously forms a bilayer, that's quite easy; but if you start thinking about the origin of molecular machines, the protein-building factories called ribosomes, it's very hard to know how these appear in the first place.

What is selection operating *on*, exactly, because we have no genes at that point—so selection on *what*, exactly? Or is it just some sort of spontaneous, self-assembly? In which case, we have the obvious questions, *Why **this** structure, exactly; why does it happen to do that job rather than another structure?*

HB: Right; and there's also the question about the scale. In your book, you gave some anecdotal notion of this by saying, I think, that there were 13 million ribosomes in one cell, all of which are efficiently producing proteins—that's just mind-boggling.

NL: Yes, so there are 13 million of them inside one cell—and you can't even see the cell—but at the same time, if you go down to the level of atoms, then these are massive, molecular machines with 60 or 70 moving parts—and yes, they can make an entire protein in a minute or something like that, so they're extraordinary things.

So these are the *real* problems for the origin of life that have barely been touched on yet. We have no idea how long that will take or how likely that is, but one thing you can say for certain is that if you don't start from an environment which provides the building blocks for all of that and then somehow focuses it in that particular direction, you will never get you there.

So this chemistry must happen quickly and it must happen well. The fact that we can make it happen a little bit in the lab is an encouraging start, but we've got a small reactor and atmospheric pressure in a glass jar and we're talking about simulating reactions which should have been going on across the sea floor in vents that can be 60 metres tall at high pressures where hydrogen dissolves properly.

You don't know what the missing parts are. These are geologically, chemically complex systems, all we can do is say, "*Well, modern cells that live there use these iron-sulphur minerals*", for example, or, "*They use magnesium to do this*"; and so we try to use those same things—we try to reconfigure them from the same ingredients that cells use—not the things that might work best, but the things that cells actually use.

Questions for Discussion:

1. Might there be some sort of chemical process which happens over very long time scales? What would such a process need to involve?

2. To what extent do Nick's views on evolutionary processes correspond with the notion of "punctuated equilibrium" made famous by Stephen Jay Gould?

V. Synoptic Justifications

Philosophers wanted

HB: Well, that all makes sense to me given your constraints; and I don't pretend to have any knowledge of the differences in scale that arise when you have one of these things that's massive with high pressures and all the rest, but one thing that *did* occur to me when I saw this little tabletop thing that you had is, *Why don't they have big, biological experiments like they have big physics experiments?*

NL: We can't agree with each other. I mean, everything I'm saying now is a point of view which was pioneered by Mike Russell and Bill Martin and there's probably 20 of us in the world actively working on these kinds of questions, specifically to do with the origin of life in alkaline, hydrothermal vents.

   The majority of people working on the origin of life are coming at it from very different places; and who's to say who's right and who's wrong? I think we're right, obviously, or I wouldn't be here, but they think they're right and they may well be right and science is not at the stage yet where we can build a CERN or something where everyone agrees on what the question is and how exactly to test it.

HB: OK, but I mean 20 people for this approach? I'm sure you're right, I'm not arguing with you, but that seems odd to me. I can certainly imagine that there's a spectrum of different views, but why is that, do you think? Is it because people are primarily transfixed with genetic mechanisms, or that the chemists aren't thinking biologically, as you mentioned earlier? What's going on? Why are there only 20 people, more or less, working on this?

NL: I think it's precisely that the chemists *don't* think biologically. They think in terms of what chemistry is going to work, and they're not too worried about the next steps—they're not too worried about how a ribosome self-assembles, say—they're interested in whether or not they can do the chemistry that can make the nucleotide building blocks; and that's enough for one person's lifetime in science—it's a difficult question.

Then there are the astrophysicists who are interested in cosmic chemistry. It's a fact that meteorites are often chock-full of organic carbon that was being delivered to the early earth—there's no doubt about that. But the question is, *Is it meaningful?* Did it deliver something that was *needed* for the origins of life or was it incidental?

My guess is that it was largely incidental, but why should people believe me beyond the power of an argument? Science is not about the power of argument: science is about the evidence base for the argument.

Arguments are perceived as dangerous by people—I think too much. Although a PhD technically means a Doctor of Philosophy, I think there's not enough philosophy in science these days, actually.

One of the great things about having written books, and thought about this broader picture, is that I now have what I hope is a very solidly grounded picture of how these things fit together across a large scale; and this has led me to conclude that the answer has to be in a certain well-defined area.

Maybe I'm wrong in the details, but I'm convinced that it has to be there somewhere. I find that that's true of a lot of these types of questions: people who are in a particular field and are not taking that broader, synoptic view will tend to compare different ideas solely on their own perceived merits or demerits: "*This idea has these strengths and it has those weaknesses, and that one has these strengths and those weaknesses*".

And then you might compare them. So a review of origin of life chemistry will cover hydrothermal vents and might say something like, "*Okay, maybe there's something about the structure. It's difficult;*

people have tried to make hydrogen react with CO$_2$ and yes, you get trace amounts of this".

And the conclusion would likely be something like, "*Well, it's an interesting idea, it's experimentally tractable and some people are working on it; but there's been a lot more weight of evidence in this other idea*"; so an objective, balanced, overall review would tend to put less weight on hydrothermal vents than it would on other environments.

HB: Well, in addition to this key synoptic feature you're alluding to, it should also be mentioned that I suspect part of that is because not nearly as much work has been done on hydrothermal vents either— it's a chicken and egg type of situation.

NL: Absolutely, that's a large part of it. And also because it's not easy for these reactions to happen, because nobody's looked at the structure of cells and tried to work out whether or not they can be driven to occur.

But as I said, the reason that I think the answer lies there has got nothing to do with the quality of the science in the other fields, it's got everything to do with the philosophy that's associated with the very notion of life itself, leading me to ask, "*Well, these proton gradients across barriers are so fundamental to life that they **must** have arisen early; and they **must** have happened in an inorganic context originally, so **what** could that inorganic context possibly be doing?*"—a question to which there's only a very limited number of answers.

So whether or not it works for *me* in the lab, I'm confident that it will work—that having a broader view can funnel you down into a particular question; and even if the evidence for it at the moment is weak, which it is, I think 10 years down the line, the evidence for it will be strong. And if it's not, then I'll be out of a job.

HB: So—just to hopefully recapitulate what you were saying—if you come at it from the idea that this proton-motive force is the most fundamental thing, or at least a very fundamental thing, then you're necessarily inclined to look at mechanisms that can generate life that naturally have that, first and foremost.

That's your guiding philosophical picture, which doesn't just arrive randomly. It came from looking around and saying, "*My goodness, all these cells have this as an essential aspect to them*".

NL: Which is something we've only realized over the last couple of decades, and the more cells that we discover living by very different means in very different environments, turn out to have exactly same mechanism.

HB: Such as these so-called "extremophiles".

NL: Exactly—that's right, yes.

So, the strength of that argument has become enormously stronger over the last decade: that this is a fundamental aspect of how cells work.

None of this is to gainsay the importance of DNA and information, but it's just to say that all of that operates within the context of a cell, and cells operate within the context of energy fluxes across a membrane.

And that membrane and the proteins involved are so complex that most people simply put it to one side and say, "*It cannot be ancient because it's too complex*", rather than saying, "*Okay, so what* are *they actually doing and can we conceive of intermediate steps on the way to that happening?*" That's essentially what I've been doing.

So, this overarching synoptic approach makes me feel quite confident that there has to be something in it, and it's just a case of how one gets the funding to do the experiments and how one physically sets up the experiments—those are the kinds of questions that I wrestle with on a daily basis.

Questions for Discussion:

1. Do you agree with Nick's statement, "Science is not about the power of argument: science is about the evidence base for the argument"? How, exactly, should "evidence" be defined in this context?

*2. Should all scientists be required to take a course or two in philosophy of science as part of their training? (Readers interested in this issue are referred to Chapter 5 of **Inflated Expectations: A Cosmological Tale** with Princeton University physicist Paul Steinhardt and Chapters 3–5 of **Constructing Our World: The Brain's-Eye View** with Northeastern University neuroscientist Lisa Feldman Barrett.)*

VI. Becoming Complex

From prokaryotes to eukaryotes

HB: So at the risk of beating this to death, my sense is that you're focused on the fundamental importance of this proton-motive force—that's the thing that is deeply tied to your entire philosophical approach as a core feature of life—and you have to find a way to get that to happen. You're focused on these hydrothermal vents because you think that they are a prime candidate for somehow generating this key cellular mechanism from some related inorganic process—and tautologically this process has to be inorganic to start with, of course, since we're dealing with the origin of life—but even if somehow these hydrothermal vents don't work at the end of the day, and you have to look elsewhere, that hardly invalidates your key motivation and overall approach.

But then there's this second aspect that we talked about at the very beginning: once things start and you've got cells with membranes and energy fluxes and all of that, how do we go beyond that and get the complexity of life—not only the difference between Bacteria and Archaea, but also—eventually and rather more significantly in my view—the appearance of Eukarya like us—all of these eukaryotic cells with a nucleus inside, and all the wonderful biological complexity that springs from that.

NL: Well, that does stem, in fact, from these conditions at the origin of life.

Let's say, for the sake of argument, that this what I've been telling you is correct: that the reason that all cells use proton gradients across membranes is, in the end, to drive the reaction between carbon dioxide and hydrogen, to make organics and to grow—it's that fundamental.

These vents are produced by a reaction between rock and water and they're found on any wet, rocky planet and icy moons in the solar system—Enceladus and Europa and places like that—even Mars has these hydrogen-producing, serpentinization reactions going on. Carbon dioxide is very common in atmospheres of bodies in our solar system, so there are plenty of occasions where these reactions can occur.

More generally, there would be billions of these sorts of exoplanets in the Milky Way alone, so you're going to have repeated on billions of occasions these occasions where you have hydrothermal systems rich in hydrogen, and atmospheric conditions rich in CO_2 and the barrier to them reacting is lowered by the structure of the system—that you've got a different pH across this barrier.

So, if these ideas are correct, then it immediately explains why Bacteria and Archaea independently have the same properties: that they're using the membrane to react hydrogen and carbon dioxide to make organic molecules.

Even plants are reacting hydrogen with carbon dioxide: it doesn't look like it, because what they are doing is stripping the hydrogen from water—they're using the energy of the sun to split water, grab the hydrogen and do what these cells always did—and the waste is the oxygen that gets washed out into the atmosphere.

Again, this is a fundamental principle of what life is doing: it's using carbon dioxide and attaching hydrogen onto it. It's getting that hydrogen from wherever it can now but in the beginning it was just bubbling out of the ground.

So you have a cell, which is basically growing because of proton gradients across membranes, and these same constraints should apply to life anywhere else.

It's imaginable that life could operate in a different way, but because these conditions are so common—indeed ubiquitous—it's likely that you would have similar constraints on life arising on other planets for similar reasons.

And then, you're going to get stuck, because the problem with respiring across the membrane is that you run into two sets of constraints.

One of them is just a geometric issue related to surface area. If you're breathing across your skin, then the amount of energy you can get depends on the *surface area* of your skin, whereas the amount of proteins you need to make to be an active, working cell depends on the *volume*; and as you become a larger and larger cell the surface area falls in relation to the volume.

So, there are constraints there. Now, why can't you "internalize" the situation and get out of those constraints, as we do with our mitochondria?

And the answer there seems to be that what you have are membranes with a charge across the membrane—the proton gradient is a charge, and it's a very strong charge. If you get down to the level of the membrane itself, it's about 30 million volts per meter—that's the field strength that you would experience if you were the size of a molecule down there, which is equivalent to a bolt of lightning.

In all of our cells, then, the membranes have a charge across them equivalent to a bolt of lightning; and that's what's really keeping us alive—it's gone back generation after generation, right back to the very origins of life. It's a thrilling thought, really.

So how do you keep a charge with that intensity from doing damage?

Well, the answer seems to be, *By responding in a, kind of, thermostat-type way to any changes in the intensity of the charge or the conditions that generate or dissipate the charge*—and we have to react within minutes, usually.

So if you put a plastic bag over your head, you're going to be dead in a couple of minutes. If you had a partial plastic bag over your head, maybe you would survive—but if you survive, it's because you're adapting to the changing conditions; and again, you had better do it quickly.

All diving mammals adapt very quickly to having oxygen cut off, and it seems to take genes right on site—it's as if there's a requirement for a sort of "local management centre" to organize on the ground. You could see it in military terms: a local commander on the ground responding to changes in the immediate circumstances.

So this seems to be the real problem: we have inside ourselves these mitochondria—they started out as Bacteria, they had whole genomes, bacterial-sized genomes, of probably four or five thousand genes; and they got whittled away over two billion years and down to just a handful of genes left.

But we all have that same handful of genes left: we do, plants do, fungi do—basically *all* of complex life has *two* genomes, not one genome: the massive, nuclear genome, which codes for all the complexity, but that massive genome is supported by these tiny genomes in the mitochondria.

And if you were to strip those genomes out, then effectively, the cell would lose control over its own energy and be unable to maintain its own nucleus.

So, the hypothesis is that you cannot have complex life with massive genomes unless it's supported by these tiny, pared-back, bioenergenetic genomes.

HB: OK, so if I understand the story correctly, the idea is that at one point some bacteria was swallowed, as it were, by an archaean cell and eventually became mitochondria, thereby creating a new type of cell—a eukaryotic cell—that allowed all of these other complex structures to develop.

I can see that there are energy considerations and so forth, but one obvious question is, *How often did this happen?*

My recollection is that, according to theoretical arguments, this only happened once around 2.5 billion years in. I can sort of see how all sorts of interesting developments can take place after that point, but that certainly strikes me as curious, and I'm left wondering, *How did that happen?* And, relatedly, *Why only once?* Maybe, somehow, those sorts of processes are happening all over the place.

NL: Well, I think we need to take a step back and ask, *What are we trying to explain here?* Most people's immediate perception is exactly that, "*It happened once, that's outrageous, it cannot possibly be true, it's too much of a freak fluke*".

HB: I didn't say it couldn't be true; it just strikes me as requiring some sort of an explanation.

NL: Well, as human beings, we back away from that proposition; and I think it means that we tend not to look at the evidence in an even-handed way.

So what's the evidence? I have to say that it's weak, but what do we have? Well, what we see is that we are all very closely related to the plants outside our windows at the level of our cells, at the level of our genes in thousands upon thousands of ways. We undoubtedly share a common ancestor—and, by definition, that common ancestor can only have arisen once.

The first thing, then, is that we look at all this complex life around the world and we find that it all shares that same cell structure—which is odd, because what I would expect from first principles, would be that plants, for example, would have arisen from photosynthetic bacteria and I would have arisen from some kind of bacterium that went around eating things and a fungi would have a arisen from another type of bacteria that went around dissolving the matrix that it lived in.

But that's not what happened: we all go back to a single, common ancestor—and as I say, by definition, it arose once, and it's very difficult to track what happened before that ancestor.

So, we share all of this complexity with all these other things and Bacteria don't and Archaea don't and we've spent a lot of time looking in all kinds of environments. It wouldn't be very surprising to find another type of cell which is not like either the eukaryotic cell or either of the two known prokaryotic groups.

But we've spent an awful lot of time looking and we've really not found anything that we can agree about.

HB: Wasn't there some weird thing that was found near Japan?

NL: Yes, there's one weird thing in the sea off the coast of Japan and all we've got is one section of that cell, which killed the cell, of course. So we don't have any genome, we don't know what it is.

I finished the book with a discussion of that and what it might mean, but it doesn't really take away from this problem that all complex life that we know about shares this common ancestor—which, by definition, arose once.

It arose, we think, about 2 billion years ago. So, the first thing you might say is, "*Let's scrutinize the fossil record before that; is there any evidence at all that there were other origins of complex life?*"

Now, there is plenty of bacteria in that fossil record, but there's no sign of more complex cells. Now, it may just be that they didn't fossilize very well, but we do see them later on.

From about a billion years ago, we see unequivocal eukaryotic cells—we see fungi, we see red algae—so some of these groups do fossilize just fine, it's just that we don't see them for the first 2-2.5 billion years of earth's history.

Then we *do* start seeing some things, but we're not sure what they are. They're probably eukaryotic, but we don't know. It's possible that there were multiple origins at that time of complex cells that disappeared, a lot of them, but we don't really know. So there's an uncertainty there.

And ever since then, the assumption is, "*Well, now there are these complex, eukaryotic cells that have taken over the whole world and they're going to eat everything else, so any bacteria that is striving to be a little bit bigger, a little bit more complex, is just going to be eaten before it ever has a chance to get very far*".

That is a bit of a limp assumption. It might be true, but we do know of thousands of different cells, which are a kind of intermediate level of complexity and they do just fine; but they've all evolved *down* to that level, they've all become simpler from more complex eukaryotic ancestors, rather than evolving up to that niche from simpler, bacterial ones.

So, I think what I would say is that it's a bit of a complacent assumption that it must have happened thousands or millions of times, because there's not a jot of evidence for it. All I'm trying to explain, really, is what the evidence says at face value, which is that it happened once.

Now, I think that's an extreme view; I would be shocked if it happened only once—it probably happened thousands of times, millions of times, who knows? The thing is, it only *survived* once; it only took over the world on one occasion.

So, *why?* Well, I think this takes us back to, *Why is life the way it is?*

All these cells have these particular traits; and the argument is that those traits arise because selection has switched from being about the outside world to being about the inside world: how do you survive with your bacteria population living inside you?

There are two aspects to this.

One of them is that it's really not easy for a bacterium to get inside another bacterium. We know of some examples, so we know it's possible, but it's rare. So, this is part one of the bottleneck, if you like: that it doesn't happen much.

Part two of the bottleneck is that, once it's happened you end up with an extremely unstable cell, which is likely to go wrong and likely to die.

And an argument of the book is that trying to reconcile these relationships between the cell and this population of bacteria living inside leads to a whole host of traits: that's why we end up with two sexes, why we age and die and so on. All of these traits can all be traced back to that relationship.

So perhaps there have been several other origins of complex life on earth that just never really got a footing. I hesitate to say that it's as rare as it appears to be, but what I would say with some conviction is that just having a population of bacteria on a planet is no guarantee that you're going to end up with complex life.

There's a point at which the mode of evolution shifts and it seems to require an endosymbiosis—one cell getting inside another one—for reasons that I think relate to energy and go back to the very origin of life; and that this happening, in itself, constrains the properties that complex life is going to have.

So, this is the overall thrust of the argument and it may be that it all falls down tomorrow because a single discovery could overturn a lot of this.

Questions for Discussion:

1. What part of Nick's argument would be reinforced if we were to discover a new form of bacteria on Mars tomorrow?

2. To what extent is it possible to meaningfully assess the probability of eukaryotic cells arising from the process of endosymbiosis given our current state of knowledge?

3. What might be a mechanism to explain the "whittling away" of the original bacterium genome that eventually became the smaller mitochondrial genome that Nick refers to in this chapter?

VII. Marvellous Mitochondria

Insights from a metabolic worldview

HB: So once again I'm forced to preface my comments by explicitly saying that I don't pretend to know anything about this, but it seems to me that this is the sort of thing that might be somehow modelled, trying to assess the likelihood for one sort of scenario over another.

Of course particularly in the beginning a lot would be thrown in by hand, but you could certainly imagine in principle running various models based upon these constraints and coming up with various probabilities of eukaryotic cells happening once every three billion years or something like that.

NL: You could certainly model all of that, yes. I think you would have considerable doubt about the value of the many parameters that you would chose for the model.

It would be an interesting thing to try to do, however, and I think one of the best things about modelling is not so much that it necessarily gives you an answer that you can believe, so much as it makes you think about the question in a far more rigorous way regarding what the specific parts are and how they interact with each other.

We've actually done quite a lot of modelling about questions such as the origin of two sexes: why it is that one sex passes on the mitochondria and the other one doesn't?

And, to me, there have been some revelations which come from the mathematical modelling of these questions and thinking in rigorous terms: *You've got this population of cells that are mutating; we know that they mutate, how fast do they mutate? What happens if you have a mutation rate of x? How does the host cell cope with that?*

It turns out that some very simple answers come out from that, in the end. Effectively, it's all about how you keep those mitochondria under control so that they're doing a job for *you* rather than doing their own thing. How do you select for *their* function so that they are continually being helpful?

And the answer is that you've got to mix things up. If you're a cell and you're dividing into daughter cells, what you have to make sure is that some of the daughter cells get different selections of mitochondria.

One type might be good for you, one might be bad for you, there might be a mixture in the middle, but what you're doing is that you're apportioning them out into cells that can then compete between themselves as cells—these guys have the good mitochondria, and those guys have the bad ones, so the first guys are going to win the battle.

And that's a way of selecting, at the level of mitochondria, that the population inside you has got the good mitochondria. And the only way you can do that—given that you're constantly throwing up new mutations and new "bad" mitochondria—is to sample, in one way or another—constantly passing out and allocating different types of mitochondria. And that's what's happening when you have two sexes: only one sex is passing on the mitochondria, which is a form of sampling.

They're doing it, again, in oocytes in the female germline— they're passing them out through a series of randomizing processes that mean that all the different oocytes are getting slightly different populations of mitochondria. Some of them will work well, some of them won't work well. The ones that don't work well will more likely than not be killed off, while the ones that do work well will go on to seed the next generation.

So there are insights coming from the mathematical modelling of a very simple process of generating variation by segregation, by doubling and apportioning out randomly. I think these are valuable insights, but to try and put a number on how long that's going to take is very difficult, because we don't have enough insight into what the real parameter values actually are.

HB: OK. I'd like to back up a bit to a higher level, returning to this idea of this key, overarching, recognition of the importance of appreciating the value of proton gradients across membranes. Perhaps I'm just a very easily-influenced fellow, but this strikes me as an important observation that every life form that we see around us works that way at the cellular level.

That's very suggestive to me of something very much like a principle. Now it's possible you might say that my physics background makes me "unduly principle-obsessed" or something, and I suppose it's logically possible that it's some sort of a coincidence somehow, but to me, at least it's certainly suggestive of something pretty important. After all, if you see the same thing at the heart of all life forms all over the place, it seems to me that you be rather remiss if you don't ask yourself, *Wow, that's odd, why is it that way?*

So I'm just wondering if, at the highest level—without going into all sorts of details or corroborating specific predictions with various mathematical models—or even, for that matter, going into the question of hydrothermal vents as a concrete mechanism for the generation of this mechanism—is the basic recognition of the seeming universality of this proton gradient membrane energy mechanism for life gaining wider currency amongst biologists? Do they see it somehow as a natural consequence of more fundamental principles or ideas? Or is this something that most people in the field just somehow ignore?

NL: I would say yes, it is gaining wider currency, but I think it's had a trivial effect on the overall direction of, let's say, medical research. And that's for understandable reasons, which is to say that we have tremendous processing power now and reading genes and reading variations in genes has been tremendously important and successful. And lot of the biology that I talk about in the book has come from genetics and has come from reading genomes.

HB: Of course. Maybe I didn't ask the right question; I'm not suggesting that someone who is a cancer researcher just drop everything and start doing something else.

NL: Well, actually a lot of cancer researchers are thinking in terms of metabolism now; and the field has gone, over the last 10 or 15 years or so, away from the idea that it's just mutations in genes—that it's oncogenes or tumor-suppressive genes—and people are thinking far more openly in terms of epigenetic changes, in terms of genomic stability, in terms of what's known as the "Warburg effect", that cells will switch their metabolism away from respiring oxygen—which they still do, to some extent—relying on fermentation processes instead.

I think these ideas are more mainstream than they were—and a lot of what I've written about is not specific to cancer in any particular way—but the point I'm making is that we have this tremendous ability to read the genomes of cells, even single cancer cells, but I think the field is trying to work out why the genes are being expressed in some particular way; and there's now an understanding that, underpinning that, is something about the metabolic status of the cell.

And when it comes to the metabolic status of the cell, from my point of view—since I'm convinced that the origin of the eukaryotic cell was about the relationship between two cell types, one of them was the mitochondria and one of them was the host cell, and that all this eukaryotic complexity arose in that context—the mitochondria, are, almost necessarily, right at the heart of all of that.

Now, medical research over the last couple of decades has found repeatedly, as I was finding back in the days when I was working for the medical education agency, that free radical biochemistry is central to these things.

A lot of the free radical biochemistry is coming from the mitochondria; and, actually, at the broader level, it's the mitochondria that are central to everything. This has been found repeatedly in one disease after another: that mitochondria are potentially important.

Now, a balanced medical researcher is going to say, *"Okay, well yes, they're one part of a cell, plainly, they're important: they're important for apoptosis, this controlled cell death, they're important for providing ATP, they're important for other things too, but let's not lose our bearings here, there are other cell structures too, most genes are in the nucleus, and the endoplasmic reticulum is very important as well."* So a balanced, reasonable view will naturally try not to over-emphasize the importance of mitochondria.

But this evolutionary view, which says that mitochondria are right at the centre of everything and everything that a cell does, brushes all the rest of that aside somewhat, and says, *"No, the answer really **does** lie here; and the reason that mitochondria keep cropping up is not because they're just **part** of a cell, it's because they're **absolutely central** to everything a cell is doing"*.

I don't think that attitude has been really carried across yet, but maybe it will eventually.

HB: In some of your talks, you quote Erwin Schrödinger and his approach outlined in his influential book *What is Life?*, trying to focus on the necessary aspect of everything that we call life, and my sense is that a large part of your approach is trying to answer that question in your own way.

So again, I'm not suggesting that every medical researcher should drop what she is doing, but it seems to me that this is very suggestive of a big-picture approach to the biological sciences and I'm wondering what somebody who disagrees with you would say on that level, rather than those who would argue that other things are also important—or more important in certain circumstances—which I'm sure is an argument that can be successfully made.

Again, I'm looking at things more from the perspective of, say, appreciating the necessary conditions for life writ large: considering what would be required for life, broadly defined, to occur on exoplanets say, as you mentioned earlier.

So I'm wondering what people who are also thinking about those sorts of things but disagree with you might say.

NL: I think there are fields where people really *would* disagree with me. For example, the origin of the eukaryotic cell: was it an abrupt shift in selective forces operating on it, or was it a gradual accumulation of complexity? They would have quite strong arguments against my position. And the same goes for the origin of life.

Now, for this area of medical research, I don't think that there is a strong argument against it so much as it's just ignored, because the assumption is that it's all about DNA, it's all about sequence changes.

As I say, that's now beginning to shift, but generally speaking there's not yet a widespread determination to consider matters from the point of view of there being a necessary interplay between energy and genes and structure—that cell physiology depends on that interplay of energy flow through a cell at any time, and that the epigenetic state of genes, whether they're switched on or silenced, once again depends on that energy flow.

Now, I don't think I'm saying things here which many people would disagree with, it's just not a focus: it's just not the way that people really approach the problem.

On various occasions I've tried to think through logically, *Does it point in a different direction to the direction that medical research is currently taking?* And I can't say that it really does. I can't say that the people who are studying cancer from a metabolism point of view and looking at genomic instability are doing anything other than what I would recommend that they do.

But I think there are other whole areas of medical research where that might be different. For example, we know about hereditary diseases: lots of diseases have a heritability. And that may be, very high.

So I'm making up numbers here, but let's say for schizophrenia or epilepsy, there's a heritability of 50%, which means, in the case of identical twins, if one twin has it, there's a 50% chance that the other twin would have it.

So, obviously, a lot of it is nothing to do with genes and heredity, but for the 50% that *is* to do, apparently, with genes and heredity—what is it exactly?

The way that people tend to go about these questions is by looking at what are known as genome-wide association studies: you
go across the entire genome and you look for people who've got a
genetic variant of a particular gene and you say, "*Oh, look, these two
people have this and both of them have got epilepsy or schizophrenia
or something*", so there's an association here with this variant of this
particular gene.

Now, most diseases have nothing to do with a single gene—we
know that lots of genes play a role—and so you look for all these
"SNPs," they call them (Single Nucleotide Polymorphism), and you
effectively add up the amount of heredity that these SNPs can account
for. Typically, it might add up to 2% or 3%, and so there's the thing
known as the "missing heredity," which is the other 30% or 40%
which is not explained by these genome-wide association studies.

Now the thing that frustrates me is that, very often, people will
say, "*Well, we don't have enough resolution, there weren't enough
patients in this study; and if we make a bigger study that's 10 times
as big then we're going to pick up on a whole lot more associations
that didn't show up in this one*".

But logically, that still can't account for more than a few percent
of the total associations. Other things could be heritable. For example,
environment is heritable to a degree—we have a microbiome, which
is different from what people in Japan have—it's to do with a Euro-
pean diet and so on and that is heritable, in some sense, through the
conditions in which we live. And that could, theoretically, account
for a large part of it.

But we also have mitochondrial genes and their interaction with
the nuclear background, and that also could arguably account for
that "missing heredity". The frustrating thing is that people very
rarely look: they tend to put aside the mitochondrial genes as being
trivial and unimportant housekeeping genes because there's only
38 of them against 20,000 in the nucleus, and so they conclude that
they cannot possibly be important.

So, they put it aside and don't look—and I find that a frustration,
because it's not so much that there's an argument against it, so much

as there's just an ignorance or a lack of interest in the logical possibility that something other than the gene sequence in the nucleus accounts for a large proportion of the missing, genetic, heritability of diseases, and therefore *could* give insights into it.

HB: Perhaps another way to say this is that this lack of interest is related to a lack of appreciation of function and evolutionary history rather than mere numbers and statistics. I mean, if you just look at the numbers, as you were saying, it's natural to conclude, "Well, there are only 38 over here compared to thousands over here so let's concentrate on those". But that implies a fundamental equivalence between them all. And if you look at it functionally, and you ask, *What do these things actually do?* or perhaps better still, *Why are they there at all?* it's a different matter altogether.

NL: Yes. So these are, in a sense, more philosophical questions about why the cell is the way it is and why life is the way it is, and it guides you to thinking that the interaction between the genes in the mitochondria and the genes in the nucleus has to be central to the existence of cells at all; and therefore, if it starts going wrong for whatever reason, then there will be consequences—and those consequences will be diseases: they will be the things that go wrong in our own lives.

Once again, then, this broad, synoptic view channels down and says, "*The answer to the problem is **here** somewhere.*" Maybe the details of what I've suggested are all wrong, but the answer has to lie in there somewhere: it has to be important.

Questions for Discussion:

1. To what extent is our understanding of genetics sufficient to understand human diversity? In what ways might this understanding depend on factors that can't be immediately determined from examining the DNA? (Readers interested in this issue are referred to Chapters 7–10 of **Our Human Variability** *with geneticist Stephen Scherer.)*

2. Under what circumstances can an invocation of statistical arguments actually **hinder** *our comprehension of vital underlying mechanisms at play?*

VIII. Open Questions

From the origin of life to consciousness

HB: I'd like to conclude with a question that I often ask people of a scientific disposition, which is, If I were an omniscient being and could answer any three questions that you might have, what would you ask me?

NL: Well, I think the first one would be about the origin of life: *Was it really this way? These vents for these reasons?*

The second one would be, I suppose, about eukaryotes and complexity: *Was it really about energy? Was it really about the requirement for cells interacting?*

I mean, there's a lot in these ideas, but they may be completely wrong and they may be a figment of my imagination. Science has a knack of leading you up the garden path, and it turns out that a lot of things that you became attached to and cherished are actually just wrong, so I would like to know the answer to those.

The third one, I think, is perhaps a more rounded, human question, and that's about the origin of consciousness. I wrote an earlier book called *Life Ascending: The Ten Great Inventions of Evolution*, and there's a chapter in there on consciousness.

It's not a subject that I'm an expert on at all, but it's a subject that really had to be in that book; and I've had more abusive emails from people about that chapter than anything else.

HB: Oh, really?

NL: I mean, not terribly abusive, but saying things like, *"Why are you addressing a subject you're obviously not an expert in?"*

HB: Well, you also wrote a paper once and concluded with words to the effect of, "*The quantum effects of consciousness are too important to be left up to the speculations of physicists*". I was wondering if you got any nasty responses to that as well.

NL: Not really, actually; the physicists quite engaged with that. I had emails from Stuart Kauffman and Stuart Hameroff and Roger Penrose in response to that paper, and I would stand by what I said there.

So I'm not an expert on it, but I think that biochemistry, as a discipline, has avoided the problem; and the problem is a very simple one to me, which is that we know how neurons operate—by iron gradients across membranes, funnily enough, the same mechanism.

And they depolarize: membranes will open up pores in them and sodium or potassium irons will go rushing through, depolarizing the membrane. Somehow that gives rise to feelings, somehow, the fact that this membrane just changed its physical state means that I experience love or I experience pain or hate or just bellyache.

I do believe that it's in the biochemistry, I do believe that there's something in there to explain—I don't want to resort to the idea that it's a mysterious thing and God did it. Perhaps he did, I don't know; but I'd like to think that a scientific explanation would be grounded in the chemistry that we know about.

I'm left thinking, at the end of all of this—and I've read a lot, I read a lot of books and I thought a lot about it—that someone else would have an answer to that simple question of how depolarization of a neuron gives rise to a thought or a feeling of anything at all; and I don't think anyone has ever answered that question.

I can only think of two ways into answering the question.

One is from fundamental physics, going back to the Roger Penrose view of things that there really is a missing property of matter that we don't know anything about, and if we can understand that, then we will get an insight into what consciousness is.

I find it hard to swallow that there is a fundamental property of matter which relates to the grim reality of life—pain or stomach ache or whatever it might be—and it doesn't seem to correspond to the anatomy of the brain to me either. Now, maybe it does; I don't have that as a strong view.

The alternative, though, is that it is all about the structure of the brain and it has nothing to do with the fundamental property of matter, and that, somehow, depolarization of neurons is able to give rise to a feeling and that we can find out what that is—but, in effect, selection has been operating.

Selection has basically fooled us, tricked us; and it's made us think that these feelings are real, when, in fact, they're not: they're just properties of the nervous system operating—it's an emergent property of a nervous system. Those are words that mean almost nothing to me, but that's the context in which it's seen.

So, I have no idea about consciousness. The origin of life, I can see a way through the problems; I have some difficulty with self-organization of matter, but basically, I can see a path through the whole problem that may or may not be right, but it's not beyond the wits of man to solve that problem, it's not beyond *my* wits to solve that problem.

Consciousness, I see as very different: it's beyond my wits to solve it and so far, from what I've read, it's beyond the wits of man, at the moment as well—and that's a thrilling problem for a rising generation of scientists to get their teeth into.

HB: Indeed. Is there anything that you'd like add? Is there something we've overlooked or haven't covered sufficiently?

NL: Not really. I could talk all day, but I think finishing on consciousness is good, because that's a problem that I have no personal investment in really, but I think it's one of the great, scientific problems that we still have not gotten close to solving.

HB: Well, thank you very much, Nick. It's been a pleasure.

NL: Agreed. You're very welcome.

Questions for Discussion:

1. To what extent are researchers naturally prone to regard all phenomena around them as manifestations of the same mechanisms that they are most familiar with?

2. Will we ever have a comprehensive understanding of consciousness?

Continuing the Conversation

Those interested in a more detailed understanding of Nick's thoughts on issues discussed during this conversation are strongly encouraged to read his book, *The Vital Question*, along with his other books: *Oxygen: The Molecule That Made the World*, *Power Sex, Suicide: Mitochondria and the Meaning of Life* and *Life Ascending: The Ten Great Inventions of Evolution.*

Our Human Variability

A conversation with Stephen Scherer

Introduction

More Things in DNA, Horatio...

Biology fascinates me. But as a non-expert, I'm forced to think of things in pretty simple terms. So when I hear biologists talk about evolution, adaptability and natural selection, I always find myself asking: *What's going on, exactly? What are the physical mechanisms at play?*

After all, if a species evolves through mutations of its members, then these mutations must be physically represented *somewhere*. And where else could that happen other than in our DNA, our own personal "instruction manual" of nucleotides and genes that we carry with us in every cell.

If evolution is as strong a force as we are led to believe, then, these sorts of variations must somehow be happening all around us, resulting in a world replete with manifold diversity and uniqueness that is layered upon our common humanity. Which is—to all intents and purposes—pretty well what we see when we look around and see such differences in the people on all sides of us. So far, so comprehensible.

But when the Human Genome Project announced that their DNA sequencing experiment demonstrated that we were all "99.9% identical", things took a decided turn towards the unintelligible for me, and my first reaction was one of sceptical confusion, rapidly followed by one of embarrassed withdrawal.

Like many laymen, the conclusions seemed downright perplexing to me, but who on earth was I to question the scientific consensus of thousands of expert researchers from around the world?

Stephen Scherer, on the other hand, a world-class geneticist who built an internationally renowned research program at Toronto's Hospital for Sick Children, naturally felt less inclined to be deferential to the prevailing wisdom.

> *"When the rough draft papers came out in 2000, which talked about how we're 99.9% identical, I remember thinking, 'But we're **not** identical. My brothers and I share 50% of our DNA from our parents, and we're **nothing** alike.' You could probably pick us out in a crowd, but we're really quite different."*

Let's run the numbers. For a human genome of roughly 3 billion nucleotides, that 0.1% difference results in variations of about 3.2 million of the individual nucleotides that make up the "human genome". So that's one way to look at things.

But, crucially, it's not the only way.

Many years before the Human Genome Project reached its conclusion, geneticists had also recognized that some 0.4% of the population exhibited large-scale deviations from the norm—so-called "copy number variation"—where huge chunks of DNA, often millions of nucleotides long, were either missing from their genome or present in extra copies. All of these large-scale changes were associated with serious medical conditions like autism or Down syndrome.

There were, then, it seemed two types of variation: one for the "diseased" and one for "the rest of us". It was a picture that most geneticists and molecular biologists of the time unhesitatingly accepted. But not Stephen.

> *"I have this figure that I always show the students when I teach. If you plot out the number of different types of genetic variation and divide them into single nucleotide variation and the copy number variations, you'll see that, in fact, 0.4% of the 'normal' population, the average population, carries big chromosome structural changes. Trisomy 21 is mainly associated with Down syndrome, but there are other big segments of DNA in a very small portion of the population*

that are different from each other. 0.4% of the population have these big, big changes, and we've known about that for 50 years.

"On the one hand, The Human Genome Project talked about those 3.2 million potential single-nucleotide changes that everyone is subjected to,and then on the other hand there's 0.4% of the population who experience these large-scale chromosome changes.

"And when I was teaching back in 2002, I kept thinking to myself, **'Biology favours balance. There have got to be a lot of other variants here. Why is it that we haven't seen them yet?'**

"Well, because we didn't have the tools to see them."

He didn't develop the right tools himself. But as a self-confessed "technology guy", Stephen had the presence of mind to aggressively seek out better and different techniques to see what others might have missed.

In 2003, he partnered with Craig Venter's Celera Genomics to study the DNA of chromosome 7, his primary area of expertise. Venter had pioneered a different sort of DNA sequencing technique, called "shotgun cloning", that had also been used for the Human Genome Project. Now there was a way of comparing and contrasting the two approaches.

"We published that in Science in 2003 with Craig Venter's group. Figure 1 of that paper is probably the most underplayed figure in the field of genetic variation. In Figure 1, we compared the sequence we put together with the Celera group approach with the public Human Genome Project reference sequence.

"If you look at that figure we show that there were about 167 or so sites along the chromosome that, when we compared the sequences, showed significant differences, including pieces of DNA in one that were not in the other.

"The reviewers kept saying, 'These are just technical mistakes. You guys screwed up. You made a mistake.' We knew that wasn't the case, because we had used another form of experiments to prove

that, indeed, those variations existed. But they still didn't believe us, and the editor wanted it taken out. But I said, 'You're not getting our paper unless you leave it in. The data support it.'

"Those were the first copy number variations that were identified."

So "copy number variation" again, but this time not necessarily associated with any particular condition or disease. What Stephen and his colleagues had stumbled upon was the groundbreaking possibility that large-scale, DNA copy-number variation might be nothing less than a universal human trait, a key ingredient in allowing evolutionary variability—concrete evidence, in other words, that we were far more distinctive than the Human Genome Project was telling us we were.

More work, though, needed to be done—and, once again, with cutting-edge tools.

"The real breakthrough was this technology called microarrays, which allowed us to scan for dimensional differences in the DNA sequence. What we had previously looked for were binary differences: Was it an adenine or a thymine here? Or a cytosine or a guanine there? These are single letter changes—site by site. There was really no good technology that allowed you to look for what I would call a copy number difference, where instead of having two copies, you might have three copies, or one copy, or in some cases zero copies.

"We actually used DNA from a child that was autistic as our first set of experiments. I wanted to get the most bang for my buck—I wasn't going to run just anyone's DNA—these experiments cost thousands of dollars. At any rate, we knew that this boy had about a 6-million-base-pair deletion on one of his chromosome 7's, right near the cystic fibrosis gene. We knew a lot about this.

"When we looked along his chromosome 7, starting at the beginning, there are a few blips, then you get to where his deletion is known to be, where he only has the one copy, and it drops down a bit, and afterwards it picks up again and continues along. But along the way there were all these other blips.

"It looked to be the same site on the chromosome; and we only saw them in some families and not others. That was really copy number variance. There were all these little blips along the chromosomes where there were segments of DNA of the order of 100,000 nucleotides long (an average gene is about 30,000 nucleotides), all the way up to millions of base pairs.

Further studies and meta-studies, revealed that copy number variation transcended autism entirely. It was not only common to everyone, its contribution to our uniqueness is by far the largest contributing factor, dwarfing that of single-nucleotide variations.

"We've now done a meta-analysis of all of the data that's been generated in the last 10 years with respect to copy number variation. On average, you or I would have about 45 million nucleotides of DNA in our genome (which is roughly 1% of the entire genome) that is copy number variable with respect to the standard human reference sequence. The variable factor for SNPs was 0.1%, as I said a moment ago. So that's more than ten times as much."

Ten times as much variation? How, then, is it possible that so many others could have missed it? How could The Human Genome Project— one of the largest and most comprehensive scientific collaborations in human history—have overlooked such a humongous elephant in the room?

"The Human Genome Project made a consensus sequence of what a human DNA would look like, based on a lot of individuals. In fact, I think there were 708 different donors. To come up with a consensus you have to merge them. It's like a grey picture of what a human genome would look like.

"And to do that, because there were lots of different pieces of DNA coming together from different individuals, you take the easiest explanation: you essentially force them together and come up with the most common, linear sequence.

"Just based on the design, then, you would not see these large pieces of DNA missing, because you erase that variation when you merge things. Our advantage was that we were comparing two different

DNA assemblies: the private Celera one, and the public one. And that gave us hints."

For rigorously following his sceptical hunches, Stephen now firmly occupies a place in the pantheon of the scientific establishment. But however large the personal accolades become, they are dwarfed by the fundamental change in our understanding that his research has brought, not only the extent of our genetic individuality, but also—even more importantly still—of appreciating what it means to be human.

"It seems, then, that there are all of these "normal human beings" walking around with massive chunks of DNA either added or missing. I later found out that I'm carrying a copy number variant deletion that's 800,000 nucleotides long.

"We knew about these sorts of things before, but they were always associated with disease. It turns out, however, that all of us carry lots of these chunks of DNA that are either missing in one copy, present in extra copies, or sometimes you don't have any in a gene at all.

"It's quite amazing when you think about it."

The Conversation

I. James Watson's Legacy

From The Double Helix to NRC to chromosome 7

HB: Let's start at the very beginning. I'd like to hear about how you got interested in the work that you do. You were born in Windsor, right?

SS: Yes, I was born in Windsor, Ontario, and then went to the University of Waterloo. I had three brothers who all went to the University of Windsor, so that was a big thing for me to go away for school.

HB: Was Waterloo a natural destination? You went there to study biology, I presume?

SS: Yes, biochemistry. It wasn't necessarily a natural destination. Windsor is a very blue-collar town. Back in high school, when I would meet with my guidance counsellor (who was also my hockey coach), we would just talk about hockey. No one considered that you could have a professional career in science. Pretty much everyone went to work in the factories. In my graduating class there were over 100 students, and I've heard that only two went on to finish professional school.

HB: Were you encouraged by a particular teacher to pursue higher education?

SS: No. I had always wanted to go to university. I liked the idea of going away, sitting in a library, studying and thinking. One of my friend's older brothers was at Waterloo, so we went up there to check it out. We also went to Western (in London, Ontario) and other places, but I really liked the co-op program that Waterloo offered that combined the right mix of academics with the practical aspect of finding a job. That was what attracted me to Waterloo. I was there for five years and had some really amazing work terms.

HB: Were these in hospitals? The pharmaceutical industry?

SS: For the first three work terms I worked at an agricultural station just outside of Windsor in a small town called Harrow. I worked on viruses and how they infected plants, so everything I was learning was actually relevant to what I'm doing now. We weren't directly studying DNA, but we were looking at how viruses use DNA to infect plants and things like that. The whole research environment, learning about how research plays out in discoveries, was really fascinating.

Then I had one work term in Halifax working with the RCMP. I was in the forensic labs, setting up DNA fingerprinting.

HB: Was that normal to have a work term as far away as Halifax?

SS: Employers came from all across Canada to get students from Waterloo. I got lucky. I did a phone interview and they were impressed, so I went there for four months. It was a wonderful experience.

At the time they were just setting up this DNA fingerprinting technology in Ottawa, and I was responsible for setting it up in Halifax.

HB: Really? As an undergraduate?

SS: Yes, as a third-year student. Then, because of that experience, I was able to get a job that everybody wanted, which was with the National Research Council (NRC) in Ottawa in partnership with Labatt Breweries.

They had a PhD student at the NRC working on yeast genetics—identifying the yeast genes that make the proteins that are involved in alcohol fermentation. So this PhD student was embedded in this awesome group of molecular biologists and chemists in Ottawa. It was great to get this job, not only because you got to work with some of the superstars of science, but also because you worked for Labatt, so you got a case of beer with every paycheque.

We were doing really amazing work: cloning yeast genes to find out which enzymes worked best in the fermentation process. We were using all the technologies that people were using in the biotech industry to make things like insulin; and we were doing this all on yeast. I learned a lot. I was there for two work terms, as well as the summer before I went to graduate school. I was doing things at a very high level at that time; it was just amazing.

Gerhard Herzberg, a Nobel Prize winner, was at NRC at the time too. I think he was emeritus at the time, but he still haunted the halls. I remember one day, at lunch, I was out back looking out over the Rideau River and he came to us and said, "*You guys shouldn't be out here so long. You should get back to your labs and work harder.*" Those were the kind of people we were rubbing shoulders with. It was incredible.

HB: Just to back up a little bit, you mentioned the attraction of becoming a scholar, and that it was somewhat unusual for you to go to university given your milieu. But you must have also had a predisposition towards biochemistry.

Was that something that existed from an early age? As someone with an interest in science, you might have studied physics (for example) or something else. What was it about biochemistry that particularly resonated with you?

SS: Well, if you talk to my parents, they'll tell you that I liked to do things in nature, like look for frogs and toads and that sort of thing. But all my brothers did that too.

I remember, in grade 11 biology, we had to pick a book to read. I chose *The Double Helix* by James Watson. I learned about the process of discovery—which is what that book is really about—and then also, of course, about DNA. The whole concept of a unifying molecule that underlies all of biology, that that's what natural selection and evolution is based on, really captured my attention. From there on I pretty much decided that was what I was going to do. I went through these work terms, and every single one had a DNA angle to it in some form or another.

Back in those days, when I was at NRC, people were starting to mention, in *Nature* and *Science*, this thing called the Human Genome Project: how the tools might now be available to actually map and sequence the whole human genome. I remember reading about that on my lunch breaks.

HB: This would have been the mid to late 1980s?

SS: Yes, exactly. Those papers were coming out in 1986 and 1987.

I thought, *Where can I work on this in Canada?* because I'm really a staunch Canadian nationalist and I wanted to do my studies in Canada. I could have gone to the United States, but I preferred to stay in Canada. I looked at the University of British Columbia, McGill, and the University of Toronto.

There were two guys who were both in the Department of Molecular Genetics at the University of Toronto: Lap-Chee Tsui and Huntington (Hunt) Willard, two superstars of science. They were both doing things related to the Human Genome Project. I interviewed with them and I ended up with Lap-Chee Tsui, who, among other things, is famous for having identified the gene involved in cystic fibrosis.

I was in the lab working on developing the technologies that were ultimately used for mapping chromosomes. My role in this project was to develop the technology and apply it to trying to find the cystic fibrosis gene.

The reality is that I showed up a little too late. They had mapped the gene to chromosome 7 in 1985 and I showed up in 1987. As they were moving along the chromosome, in collaboration with Francis Collins and others, Lap-Chee said, *"You should start marching in this direction, just in case."* It turned out to be the wrong direction. But, that's part of science. I was the youngest, so I was the lowest in the pecking order.

But he also said, *"We're going too slowly in the right direction. We need to develop new tools."* By the time we developed a new DNA cloning system so we could jump along the chromosome faster, they had already found the cystic fibrosis gene (that happened in 1989).

But we then used those same tools to replicate DNA in larger forms. It's like a jigsaw puzzle: if you have 1,000 pieces that are big, it's easier to put together than 10,000 small ones. We then used that technology to piece together all of chromosome 7, or about 5% of the human genome.

The very first talk I gave—I think I was in my third year of graduate school, so this must have been 1990—was in front of three Nobel Prize winners at Cold Spring Harbor, including James Watson, whose book was responsible for putting me into science at the very beginning. From very early on I was very fortunate to be exposed to this wonderful environment full of these giants of science.

Questions for Discussion:

1. Do you think that Stephen would have become a biologist even if he hadn't read James Watson's book at a formative stage of his life? Are biologists born or made?

2. To what extent do Stephen's experience serve as tangible evidence of the importance of combining theoretical and applied experiences in the development of a young scientist?

II. In the Lab

The first hints of copy number variation

HB: Tell me a little bit more about these cloning techniques that you were developing when you were in graduate school. What were you doing exactly?

SS: There are lots of different ways you can replicate DNA in an exogenous system. What happened over the years is that, if you take a piece of human DNA and want to study it, you need to have lots of copies of it because, at the time, there was no technology to study one copy at a time.

So you take it and splice it into a vector, which is another piece of DNA that allows you to replicate it in something like bacteria or yeast. That's what we call genetic engineering, which had developed through the 1970s and early 1980s. Genentech, a big company, grew out of those technologies.

But the problem was that you could only replicate small pieces of DNA. The human genome is massive. It's comprised of three billion chemical bases, or nucleotides, of information. At the time, we didn't have the technology to dissect three billion of them and put them back together, so we were forced to splice them into much larger chunks.

Yeast, *Saccharomyces cerevisiae* or baker's yeast, which is what we used, has 16 chromosomes. It's three million base pairs long, divided into 16 chromosomes. And the chromosomes are quite simple. They have what we all a centromere, a little piece that allows them to separate when the cells divide, and two telomeres, which keep the DNA molecule stable during cell division. That's it. So if you can isolate those little pieces and put them into what we call a cloning vector, then you can clone in these big pieces of human DNA, and replicate human DNA as a yeast chromosome.

At the time there was a graduate student at Washington University in St. Louis named David Burke, who was really a hero of mine. He developed this cloning system and I was one of the first to actually apply it to human DNA. In fact, I had the record at the time: I was able to clone molecules up to two million base pairs long, human into yeast. I remember him writing to me and saying, *"You broke my record!"*

HB: How do you physically do that? How do you actually separate these things? What are you actually doing in the lab?

SS: I'll tell you: it's like cooking. I spent three years doing this and I could do it because I was given unlimited funds and freedom from my supervisor, who was working on cystic fibrosis. We had a lot of money and he just said to me, *"Get this to work."*

Essentially what we did is to start with taking blood from a human; I used my own to start. I really didn't like giving blood. The woman who took my blood—she actually works for me now—took an awful lot of it. I remember asking her, *"Are you sure you need this much?"* She just replied, *"We're taking a lot."* I think it was her way of getting back at me.

Then you add a detergent. The cells lyse (disintegrate) when you add a detergent, and then you use a couple of chemical extractions, which gives you human DNA. You could do that in an hour. I do this all the time. My ten-year-old came to my lab with his friends this summer and they can do this now. It's very straightforward.

So then you have the DNA in a liquid solution. Back then, there were these, what we call, restriction enzymes, which are like little scissors that chop the DNA into pieces. So I developed a lot of techniques for that. Then, after it's been chopped up, you put the DNA into agarose, which is like a jello—it's actually cellulose. Then you run it out, separate the molecules, and cut the pieces out.

Then you take the little yeast-cloning system I talked about — that has the two ends and the centromere that replicates it— and you have to get these two things to stick together without the human DNA shearing. DNA is stable, but if you submit it to anything too rigorous it will break into pieces, and that's obviously no good.

So what I did that was different was that I figured out a way to take the DNA that was embedded in the agarose and add the other yeast systems all together and mix it up—we're talking about microlitres of DNA here, thousandths of a millilitre.

Now there's another enzyme called ligase, which was identified 20 years earlier and which pieces these together. Once you piece them together, you've got to get it back into the yeast to replicate it. So there's a method where you strip away the cell wall of the yeast and then you can get these molecules in. That was the other trick that I had to overcome.

HB: Was this all trial and error? Or did you have a sense that we know from previous experience that these enzymes would act in a particular way? How much of this was theory, and how much of this was just going into a lab and playing around with stuff?

SS: The enzymes were all known, and they got better over time. But you have to pick the right ones, and you've got to get them to work at the right concentrations. That's all tinkering and playing around. That's what science is. I always say that if you're doing cutting-edge science, 90% of your experiments should fail, which is why you need to put in an awful lot of hours to make progress.

As a graduate student, I used to go in on Saturdays because nobody would bug me. Once I had the blood DNA I could take it through the restriction enzyme step, run it through the gel, get my vector, ligate, and then I would do ten different experiments with different concentrations of enzymes. That would take until noon. Then I had to prepare the yeast. That would take three or four hours. Then around dinnertime I would mix everything together, plate it out and wait a week to see what would happen.

I would always have many different experiments going. Usually only one of them, if any, worked. Once it does, you then work backwards and say, "*This one had this concentration of enzyme A. This one had this much DNA..."* and you go through another round to try to get things even more precise.

It's vital that it works really well, because at the end of the day, for chromosome 7, we had to have 30,000 pieces of human DNA broken into pieces to reconstruct what it would look like in the cell. The human genome holds much more than that; chromosome 7 is only 5% of the human genome.

So it was just a lot of tedious work: try this, fail at that, try it again. But once it worked, it was really groundbreaking because we could then apply that not only to chromosome 7 but also to the whole genome.

Of course this sort of thing was happening in other places at the same time. Our group took an interesting approach because we were really heavily involved in the mapping of chromosome 7. The actual sequencing to determine the exact identity and order of the nucleotides that were in those molecules (as opposed to our rough mapping) was a whole different ballgame. We didn't have the funding to do that in Canada. It was mainly funded in the UK and the United States.

But there was a company that popped up, led by Craig Venter, called Celera Genomics, which used an entirely different approach than the public Human Genome Project did.

The public project was to map smaller pieces and then use what's called Sanger DNA sequencing, which is what everybody did at the time. You just needed to have a lot of machines and a lot of money.

But Craig said, *"There's no way. That will take too long."* And he really turned everything on its head by saying, *"We're just going to take that blood DNA that's sitting in liquid, chop it up using a differ-ent set of enzymes, sequence it en masse, and then use computers to reconstruct its order and identity based on how they overlap with each other."* They called this the "shotgun cloning" approach.

It was really clear to me at that time that he was going in the right direction. I talked to him and his second-in-command, Mark Adams, and they recognized the fact that they had an ally in Canada that wasn't under the influence of NIH funding.

SCHERER - IN THE LAB

For chromosome 7, we used their approach with our mapping data that we were generating, and put together the first DNA sequence of chromosome 7. There was another sequence that came out at the same time from the Human Genome Project so we could compare and contrast the two—which, in the end, was actually the best way to do it.

We published that in *Science* in 2003 with Craig Venter's group. Figure 1 of that paper is probably the most underplayed figure in the field of genetic variation. In Figure 1, we compared the sequence we put together with the Celera group approach with the public Human Genome Project reference sequence.

If you look at that figure—and I had to argue with the editors of *Science* really hard to keep that in—we show that there were, if I remember correctly, about 167 or so sites along the chromosome that, when we compared the sequences, showed significant differences, including pieces of DNA in one that were not in the other.

The reviewers kept saying, *"These are just technical mistakes. You guys screwed up. You made a mistake."* We knew that was not the case because we had used another form of experiments to prove that, indeed, those variations existed. But they still didn't believe us, and the editor wanted it taken out. But I said, *"You're not getting our paper unless you leave it in. The data support it."* Those were the first copy number variations that were identified.

HB: So the beginning of our recognition of copy number variation arose from comparing these two different approaches to gene sequencing.

SS: That was the first hint we had that sort of raised a little flag in my mind that these things might actually exist. Because we were using a computational approach, the type of variations that we were describing were very small. There was a lot known about genetic variation at the single nucleotide, or single-base level.

These ones were mainly in the hundreds or thousands of base pair size. We actually went back and used the same methods that were used in the public Human Genome Project, this so-called Sanger sequencing—

HB: Just to prove that they were there.

SS: Yes. We knew that they were there, but these were small. There was a lot known about them, but it wasn't known that there were so many. You can say, "*Well, 100 on a chromosome that has 150 million base pairs, is that really a big deal?*" But it was the first hint.

In the later 2004 paper, where we used another technology called microarrays or gene chips, the first reference we cited was that 2003 paper, because it added some evidence that what we were seeing in the later study was real.

Questions for Discussion:

1. What, if anything, can be concluded about the strengths and weaknesses of scientific peer review from this chapter?

2. To what extent does the level of current technology influence our scientific beliefs?

III. Chromosome 7

Mapping genetic markers to chromosomes

HB: Why chromosome 7? You were talking about isolating the cystic fibrosis gene earlier. What led people to that particular chromosome?

SS: We were looking at chromosome 7 because of cystic fibrosis.

HB: Right. But how did they know to look there and that that was the chromosome to be paying attention to?

SS: This is a fascinating story in itself. You should talk to Lap-Chee Tsui about this. But, in a way, I've become the custodian of the history of this.

   The first concept for deriving a map of the genome was based on mapping genetic markers that have different characteristics in different people. It's called a linkage map. This was in 1980.

   The idea was that you could use these little signposts along human DNA because they had different variations, or sequences, in different individuals. Based on Mendel's laws of inheritance, if one marker is always close to a trait you're studying—in this case, cystic fibrosis—you can assume that it's close along the chromosome, because during meiosis your germ cells develop as chromosomes that are homologous to each other: the one from your mom and the one from your dad mix together. So if you have a marker close to a specific trait, they should, more often than not, stay close to one another, rather than going to a different chromosome. And you can develop statistics around this.

   This approach, called genetic mapping, was used for Huntington's disease. It was Jim Gusella who first mapped the Huntington's disease gene to chromosome 4. Then Lap-Chee mapped the cystic fibrosis gene to chromosome 7: he had this genetic marker that was known to be on chromosome 7, and it was always segregating with cystic fibrosis.

HB: So, when you're doing DNA analysis, you take the DNA of a lot of people with cystic fibrosis and you compare it to other people, and you look at the statistics, and you're able to see that these markers are there?

SS: Exactly. Typically you do three types of experiments: you look at the people who have the disease you're studying, you look at controls who do not have the disease, and then you can look at the families of people who have the disease.

Cystic fibrosis is recessive; you actually inherit a faulty copy of each gene from mom and dad, and you have to have that double hit. Huntington's disease, meanwhile, is autosomal dominant: you only have to inherit one faulty copy, so it's a 50% chance, whereas, with cystic fibrosis and other recessive diseases, there is only a 25% chance.

Then you combine these statistics and see how these markers segregate in the parents too. For CF, the marker has to be in both. For Huntington's, it only has to be in one. So you can use little tricks of statistics to get some likelihood that a marker is linked to a disease.

In 1985, they found a marker on chromosome 7. Again, you can use these statistics to estimate, based on how often the chromosomes recombine, how far you are away. But these techniques were really crude back then. They estimated, if I remember correctly, that they were probably 30 million base pairs or so away. At the time, there were only a handful of markers on any given chromosome. No one even knew how many genes there were in a genome.

The approach used was just "walking along", finding more markers and trying to map them to the chromosome until you cross one of these places where the DNA can recombine, and then you know you're going in the right direction. This was unbelievable when you think about it. They ended up having to move along the chromosome 2 million nucleotides to find cystic fibrosis. They had no idea, though.

And that's why we had to develop methods to get bigger pieces of DNA. At the time you could move about 40,000 nucleotides at a time, if you were lucky, which would have taken forever. Then there were methods that allowed you to go 50,000, and even 100,000. This was why we wanted to develop these tools that allowed you to move along the chromosome at faster and faster speeds.

Anyway, the CF gene turned out to be on chromosome 7. Then, because we were studying chromosome 7 so closely, we became the world experts on it.

In the early days of the Human Genome Project—which is not well documented, actually—there was an international group of scientists (I think it started in the early 80s, though maybe it was the late 70s) who would get together every year at different places around the world. They would bring all their scientific papers with them in a suitcase, and they would sit around a table.

There was a chair of each chromosome. My supervisor, Lap-Chee, was the chair of chromosome 7, together with a gentleman in Germany named Karl-Heinz Grzeschik. There were people who did the X chromosome, and others who did chromosome 1, and so forth— typically those were the people who were working on the diseases on those chromosomes and had the catalogue of markers. If anyone else then found a marker on that chromosome they would come and work with you, because you had all this information and these specialized resources.

CF was really the first. We ended up doing many different diseases, including pancreatic cancer, and other developmental diseases. The first autism gene was mapped to chromosome 7. So everyone came to us. Eventually it was dubbed "Canada's chromosome", because we had an expertise there, we had added knowledge that others were seeking.

Questions for Discussion:

1. In what way does this chapter highlight the importance of statistical approaches in the biological sciences?

2. How would the presence of many different types of genes associated with a similar illness complicate the type of analysis Stephen describes here?

IV. Back to Basics

Nucleotides, DNA, chromosomes, genes and mutations

HB: Let me return for a minute to establish a few of the basics, and then we can move back towards what you have just nicely led up to a few moments ago, namely the discovery of these large-scale copy number variations, together with the implications.

But first, my understanding is that DNA consists of nucleotides, these basic elements, in this double-helix structure. And then all of this stuff is somehow divided into 23 pairs of these things called chromosomes—so our DNA is grouped into these 23 little packages, and you get half from your mom and half from your dad. And the number of nucleotides per chromosome is … roughly what?

SS: First, regarding chromosomes: there are 22 autosomes, and then there are the sex chromosomes, the X and the Y, with a male having the pair XY and a female having XX. Chromosome 1 is the biggest chromosome, which is why it's called "number 1". It has 240 million or so nucleotides. Chromosome 7 has roughly 150 million. The X chromosome is almost the exact same size as chromosome 7, so when you look at them under the microscope, those are the two that look the closest, because they're almost the same size. But, generally speaking, chromosomes 1 to 22 are based largely on the size.

HB: Okay. That I get, more or less. But what I have an embarrassingly hard time understanding is the following: you've got these nucleotides and you've got DNA, which has this double helix structure, and it's grouped into pairs of these different chromosomes of varying sizes, but what is a gene exactly?

This term is part of common parlance. We talk about how there is a gene for this and a gene for that. But what does that really mean? How do you actually *define* a gene? And how does that link up to the protein that is produced?

SS: The chromosomes are the molecules that contain DNA. So every chromosome is made of DNA. Along that strand of DNA you actually have segments of DNA that encode proteins, and the intermediary between DNA and proteins is RNA. After lots of work on the Human Genome Project we know that about 1.6% of all of the 3 billion chemical bases, or nucleotides, actually encode genes. The other 98% or so is there for structural purposes, to keep the chromosomes together, and to turn genes on and off.

But there are some signposts, or hallmarks, of what gene sequences look like. This is work that Francis Crick, Sydney Brenner, and others did. They actually looked at the DNA sequence to find out what the combinations of the triplet code were to make the proteins. There are three letters that come together to make an amino acid. If you work backwards there are certain signposts along the DNA sequence that tell you that that sequence actually makes what are called the exons, which are spliced into RNA to make amino acids that combine to make proteins. So it's a genetic code.

HB: So what does 'gene' mean, exactly? I think I know what a protein is (although perhaps not), but how do you classify a gene, precisely?

SS: This is where, for eukaryotic systems like humans, it becomes very complex. In fact, the field argued about what the definition of a gene was for a long time. The classical definition is that it's a piece of DNA that can be mutated—if you have a mutation it leads to a trait, or what we call a phenotype.

HB: So it's a backwards definition.

SS: That's a backwards definition. Then we went through the Human Genome Project and we were able to find out more about the physical characteristics of the genes. So, to get even more detailed, genes are made up of exons and introns. Exons are the pieces of RNA that are spliced into protein, and introns are the pieces that are spliced out and just degraded in the cell.

For example, there are some genes in the genome, a very small number, that have only a single exon. An average-size gene has 10 exons. I'll jump way ahead and tell you that there are about 25,000 protein-coding genes. There are some genes that have well over 100 exons. The biggest gene in the genome has something like 300 exons. There are many different forms of these combinations of exons that come together. That's what adds to the protein complexity.

HB: So I need a protein, because if I don't have a protein coming out at the end of the day then, according to this definition, it's not a gene. Is that right?

SS: Yes.

HB: So at some level, it's defined as some process involving exons and so forth that led eventually to a protein. And this protein then goes off and has all sorts of different mechanistic effects, which may or may not lead to diseases, or the inhibition of diseases, or positive or negative things happening, and so forth. And I can trace causally what's going on with this protein. Is that fair?

SS: Well, there are roughly 25,000 genes; and each gene, on average, can make about 10 different types of proteins depending on how these exons splice in and splice out. That's perhaps too much information. But these proteins are grouped into different categories.

There are enzymes, things like amylases, which degrade the starch you eat. There are structural proteins, like collagen, that are involved in your skin and muscles and things like that. And there are other proteins that are involved in your brain cells, your neurons, how they come together and communicate with each other. Everything in your body is made of proteins, but the code that tells those proteins how to be made, and when to be made, is the DNA. The intermediary between the two is the RNA.

HB: And we call "the gene" that thing that leads to this protein that has all these various different effects.

SS: Yes. And the genome is all of the DNA. So in every cell in your body—an average human has something like 60 trillion cells in their body—the DNA in every cell is the same. That's your genome. The DNA makes the genes. You can think of the genome as an orchestra: during development, different sections of the orchestra come on or off. Those are the genes.

There are some genes that are only expressed in the pancreas. There are some genes that are only expressed in the brain. There are other genes that have different forms that are expressed in different tissues. The cystic fibrosis gene, for example, is involved in lung development and it's also involved in pancreas development, and some other places too. That's one of the tricks: we have to determine that.

To come back to the association of genes with disease, you might expect that if a gene is making a protein that's only turned on and active in brain development or brain activity, a deletion of that gene that gave you only one copy instead of two would be associated with a brain disorder, because that's where the outcome is: it's only turned on in the brain. And that turns out to pretty much hold true.

HB: I wanted to go into some detail for a couple of reasons. The first is that I'm always more comfortable when I understand what we're actually talking about, and I never really understood what the word 'gene' meant, other than 'stuff we inherit'. I didn't fully appreciate how the mechanistic aspect works, which, I think, is really important when one starts talking about mutations and what's actually happening under mutation.

The other reason is that, at the end of one of your review papers ("The clinical context of copy number variation in the human genome"), you talk about how it might be advisable to change our perspective to a more genomic perspective, rather than a genetic-oriented perspective. So I naturally wanted to try to understand what that means.

SS: You're actually making me think of a good way to explain this.

I really like this classical definition of a gene, where we talk about a change that leads to a specific measurable trait or phenotype, because it's based on model organisms like yeast, as we talked about earlier.

That is, we ascribe a genetic component to something by looking at the outcome that occurs if you randomly mutate a piece of DNA—such as an ability to better ferment alcohol, or in a human it could lead to a brain disorder, or cystic fibrosis, or something like that.

There are 6,000 well-characterized genetic diseases in humans, but there are 25,000 genes. I think if you mutate any gene you're going to have some detectable outcome, if you know what to look for.

I like to say we're really all mutants. We all have different genetic variations. Even the most healthy, "normal" individual is carrying collections of his or her own DNA mutations, but we just haven't detected what the effect might be. But that's what natural selection is. These mutations occur. In some cases they add advantage, while in other cases they are deleterious in some form or another.

Questions for Discussion:

1. What does Howard mean when he refers to "a backwards definition"? What other sort of definition might he have had in mind when he asked his question?

2. How might the number of both "genetic diseases" and "known genes" change over time and why?

3. What do you think the evolutionary advantage of having 23 pairs of chromosomes might be? How might a "chromosome" be defined rigorously from an individual nucleotide perspective?

V. Revolutionary Stirrings

Detecting copy number variation

HB: OK, let's return now to where you left off some time ago. We were talking about our level of understanding of the human genome by 2003 or 2004. My understanding is that what the Human Genome Project was mapping was all these little nucleotides in your "standard" human, whatever that means exactly.

And what you and your collaborators had come to appreciate is that the amount of variance in the genome from one person to another was not only much larger than we had been led to believe, but it was also quite different in kind, insofar as it wasn't just the change or mutation in these individual nucleotides; it was actually whole strings and all sorts of different things. Is that a fair way to put it?

SS: Yes, absolutely. The Human Genome Project, the public endeavour, made a consensus sequence of what a human DNA would look like, based on a lot of individuals. In fact, I think there were 708 different donors. To come up with a consensus you have to merge them. It's like a grey picture of what a human genome would look like.

And to do that, because there were lots of different pieces of DNA coming together from different individuals, you take the easiest explanation: you essentially force them together and come up with the most common, linear sequence.

Just based on the design of how they went about doing things, then, you would not see these large pieces of DNA missing, because you erase that variation when you merge things. Our advantage was that we were comparing two different DNA assemblies: the private Celera one, and the public one. And that gave us hints.

But the real breakthrough was this technology called microarrays, which allowed us to scan for dimensional differences in the DNA sequence. What we had previously looked for were binary differences: Was it an adenine or a thymine here? Or a cytosine or a guanine there? These are single letter changes—site by site. There was really no good technology that allowed you to look for what I would call a copy number difference, where instead of having two copies, you might have three copies, or one copy, or in some cases zero copies.

A few groups developed this technology called microarrays, and we were probably the first in Canada to get access to these microarrays from a company in Texas called Spectral Genomics (this was back in 2001 or so). They had pieces of DNA, every million base pairs, spaced along a little glass slide the size of a dime, which was something called a "one-million-base-pair chip". It was this big thing. It cost several thousand dollars to do each experiment.

Essentially what you have on a microscope slide is a scaffold of DNA sequences across the consensus human genome. Then if you take my blood DNA, or yours, and hybridize it, check to see what the differences are (you're looking for intensity differences in how they come together) you can actually scan to see if you have the same copy number—say, two copies, or one copy—because you're comparing it to another reference. It can get a little bit technical, but it's really that simple.

We were the first to get this technology. This has been a recurrent theme in my career: we always push hard to be the first to get the new technologies. It's obvious. With a new technology you can see things that others don't. It's pretty easy to make discoveries if you're seeing things for the first time. Moreover, we had our eyes open, because we weren't fixed on forcing the sequences into one consensus sequence.

HB: And how much statistical analysis do you have to do with all this stuff? Is that somehow embedded in the technology, or do you have to then take that information and run statistics on it?

SS: Those were really dirty experiments; and most of them failed. This was a real problem. There was so much statistical noise, because we had roughly 3,000 sites along the human genome. And every time you hybridize DNA using this nanotechnology—we're talking very small volumes—you get all kinds of noise. So we used a lot of statistics to smooth out that noise.

In fact, what happened was that one of my postdoctoral fellows (Lars Feuk) came back with this experiment where he was showing me all the statistical plots of these different sites along the genome.

We actually used DNA from a child that was autistic as our first set of experiments. I wanted to get the most bang for my buck—I wasn't going to run just anyone's DNA—these experiments cost thousands of dollars. Anyway, we knew that this boy had about a 6-million-base-pair deletion on one of his chromosome 7's, right near the cystic fibrosis gene. We knew a lot about this.

So when we looked along his chromosome 7, starting at the beginning, there are a few blips, then you get to where his deletion is known to be—where he only has the one copy—and it drops down a bit, and afterwards it picks up again and continues along. But along the way there were all these other blips.

I remember Lars saying, "*Okay. We got it working. We can see the deletion.*"

And I replied, "*Well, we knew that already. What the hell are all these other things down here?*"

He said, "*Well, you know, I see this coming up in a few of my other experiments,*" and he pulled out some more plots.

"*That's interesting,*" I said, "*It's coming up on the same site on the chromosome, but in different places.*"

"*Well,*" he told me, "*It probably just means that there's some technical noise there.*"

And I said, "*I don't think so. Go back and look at our chromosome 7 paper from last year where we used a different technique.*"

It looked to be the same site on the chromosome, and we only saw them in some families and not others. That was really copy number variance. There were all these little blips along the chromosomes where there were segments of DNA on the order of 100,000 nucleotides long (an average gene is about 30,000 nucleotides) all the way up to millions of base pairs.

It seems, then, that there are all of these "normal human beings" walking around with massive chunks of DNA either added or missing. I later found out that I'm carrying a copy number variant deletion that's 800,000 nucleotides long.

We knew about these sorts of things before, but they were always associated with disease. It turns out, however, that all of us carry lots of these chunks of DNA that are either missing in one copy, present in extra copies, or sometimes you don't have any in a gene at all. It's quite amazing when you think about it.

HB: When you first started out, you used DNA from an autistic child as you've said, but then later, I'm guessing, you started generalizing further in an effort to explore this basic human variation. How did that all happen?

SS: Well, it was a few things. We were limited in two ways. The first issue was that it was technically hard for the company to make these chips, because they had to actually use a printing process, believe it or not, where they stamp these things on. It really was messy science.

And, secondly, they were very expensive. I was so fortunate; I received a philanthropic donation from a very wise woman who gave us $100,000 at the time. She had a child who was autistic.

HB: Can we say her name?

SS: Her name was Louise Morgan. She has since passed away. I remember having lunch with her and I was pretty sure she was going to give the donation, but she had to do her due diligence.

To get access to the Celera data, we had to pay quite a large sum of money, something like $50,000. Although we were collaborating with the Venter group, he had a board of directors and so forth and we still had to pay some money to get it. Then, to do the first 20 experiments using these microchips cost quite a bit more—something like $2,000 per experiment, if I remember correctly.

The $100,000 that she eventually gave us would cover all of this.

HB: Pretty important.

SS: It was very important, and I couldn't be certain it was coming. I had actually made the decision that, even though I didn't really have the money, I was going to underwrite it myself. I knew it was big. As I said, when you have access to new data, it leads to new things. I suspected all along that these CNVs (copy number variants) would be involved in autism, because we had other clues that chromosome abnormalities were involved.

Anyway, it was an issue of getting the technology, having enough money to pay for it, and then getting it to work—because almost all of our experiments were really lousy. I remember once getting a whole shipment of these arrays that cost $30,000 and not one of them worked.

We were able to get more funding in place early on; and in the original 2004 paper that first described genome-wide copy number variation, we ended up combining our data with Charles Lee, who was at Harvard at the time. He was a Canadian from Edmonton. Well, he's actually Korean, but he went to school in Edmonton, so we knew of each other and came together.

HB: There's a Canadian theme through all of this. Virtually everyone you mentioned seems to have a Canadian link. You're letting your national colours show, as you warned me that you would before we started filming.

SS: It's really interesting. The guy who was down at Baylor College of Medicine in Houston making the arrays was Canadian too. He trained at the University of Guelph. His name is Mansoor Mohammed and he actually brought Charles and me together.

To publish the first paper, we knew the reviewers were going to kill us on statistics, because I had 25 or so genomes done, of which many were from autistic individuals.

But we were now investigating the general population, and combining the two is taboo. Meanwhile, Charles had about the same number done. But for us, as young scientists, these were massive budgets. This was our whole lab budget. So he said, *"To really convince the reviewers, we've got to bring our data together."*

Meanwhile, Michael Wigler, was at Cold Spring Harbor Laboratory. That is where James Watson (whom I mentioned earlier) works, and it's a very wealthy lab. They published a similar observation in *Science* in essentially the same week, and they had only 20 genomes.

Anyway Charles and I came together, each with over 20 genomes, and put our data together. We were a little bit younger then, and we figured we would need that to get it through the peer review.

There was a lot of hedging bets that it would work. But, as I said earlier, in science you need to do incremental science, but you also have to do hard science. In 'hard' science your experiments mainly fail; and that's what we were doing in those days. Then, to convince someone, you have to get your numbers up. That's why Charles and I came together.

Questions for Discussion:

1. What lessons can be learned from Stephen's description of the way the Human Genome Project was deliberately designed to disregard differences to produce a "common linear sequence"?

2. What do you think Stephen means, exactly, when he distinguishes between "incremental science" and "hard science"?

3. To what extent do you think non-scientists appreciate Stephen's remark that "in hard science your experiences mainly fail".

VI. Going Global

Large-scale variation for all

SS: Before we published the 2004 paper—which included data from autistic patients—we started to plan how we might see how much of this copy number variation there was along the chromosomes in worldwide populations. And in the fall of 2004, I was able to convince the Canadian government to fund this.

We came together with another champion at the Sanger Institute, which is the world's largest genome institute in the United Kingdom, funded by Wellcome Trust—they sequenced 30% of the genome for the Human Genome Project.

There was a really strong molecular biologist there named Nigel Carter, who actually wrote the "News and Views" commentary in *Nature* describing Wigler's paper and our paper. He knew this was important; and it was easy to convince him that we needed his input to put this multi-million-dollar copy number variation project together.

Then Nigel brought in Matt Hurles, and together we designed what would be the first copy number variation map of the genome. We ended up finding out that, when you look at world wide populations, roughly 12% of any DNA molecule along the chromosome could be copy number variable.

HB: My sense is that 12% is a huge number relatively speaking—completely at odds with what the prevailing belief was at the time, right?

SS: Yes. So it's important to step back a bit to give you an essence of the numbers.

If you think of the amount of variations for individual nucleotides, the single-letter changes of base pairs: as I said earlier, after the first genomes were sequenced we found out that there are typically about 3.2 million of these for everyone (they're called SNPs —pronounced 'snips'—or single-nucleotide polymorphisms). You have 3.2 million. Some are common to the 3.2 million that I have, and some are unique to you.

HB: So these 3.2 million are the deviations from the "normal", whatever that is.

SS: Exactly: deviations from that reference public genome sequence. These reflect your ancestry: where you come from, and the mutations that your ancestors were exposed to. This is 0.1%.

That was the big message of the genome project: we're 99.9% identical. Well, they were right, because that's all they could see at that time. When we did the first copy number variation map in 2006, where we weren't just looking at the single-nucleotide variance; we were looking at gene copy number differences—

HB: Larger scale structure.

SS: Yes. Larger scale structural changes.

In the end, we went back and looked at Craig Venter's genome that was fully sequenced by his group, and we did the annotation. It turns out that all of us actually have well over 40 million nucleotides involved in copy number variation. That's more than 10 times the diversity than we previously thought, which means that that 0.1% of difference I spoke of before becomes more like 1%. In fact, it's probably even somewhat more than that.

We've now done a meta-analysis of all of the data that's been generated in the last 10 years with respect to copy number variation. On average, you or I would have about 45 million nucleotides of DNA in our genome (which is roughly 1% of the entire genome) that is copy number variable with respect to the standard human reference sequence. The variable factor for SNPs was 0.1%, as I said a moment ago. So that's more than 10 times as much. In 2004, everybody was doing experiments without even knowing about this massive amount of variation. They were making mistakes, misinterpreting, under-interpreting.

Now we've got a lot of data, which allowed us to go back and do the meta-analysis. If you look at all of the chromosomes in the worldwide population, on average, 12% of the DNA (in terms of genes) can be copy number variable, but the remaining 88% or so is very copy number stable. That 88% only becomes copy number variable in disease situations.

And that's the way evolution works. The copy number variable regions occur where the gene may be more "relaxed", if you will—it doesn't really matter so much, it doesn't lead to a negative outcome. That's where natural selection and evolution work. If you get five copies of a certain gene, for example, maybe that allows you to digest a particular food that you're exposed to in a better way, which gives you a selective advantage.

We now know the specific genes that are allowed by nature to be copy number variable and the ones that are under constraint not to be copy number variable. We've got that map information.

Now when you take all these disease genes that have been identified, the ones that are associated with disease tend to map into those regions that are copy number stable in the general population.

HB: Those are things that go wrong.

SS: Yes. You only have one copy, and it goes wrong.

So it's amazing. In the last 10 years we've developed a whole new way of looking at disease. If you map that information back to how genes are turned on and off in different tissues, you can predict, for instance, that if you have only one copy of this gene and it's turned on mainly in the brain, then it might lead to a brain disorder.

And that's exactly what we saw when we looked at individuals who had autism spectrum disorder and their parents. In our first set of experiments (I think the first paper was in 2007) we used essentially the same microarrays we used for the copy number variation discovery. We looked at roughly 1,000 families in an international consortium.

And we found, in 7% of cases, the children who had autism spectrum disorder had a new copy number variant that was not found in the parents. This was what we call a spontaneous change, and it was affecting the gene that was involved with brain development, specifically how the neurons communicate with each other.

HB: OK. I have to pause for a second, because you've said a whole bunch of things. I want to get to autism very soon, but first I'd like to discuss this evolutionary aspect you just mentioned, because I'm trying to get a clear handle on that.

Questions for Discussion:

1. *What does Stephen mean, exactly, when he refers to some genes as being potentially "more relaxed" with respect to copy number variation?*

2. *Might the property of "copy number variability" itself be variable?*

VII. Variability and Evolution

Appreciating biological variability

HB: So there is this 12% variation, on average, in terms of copy number variability. But the principal point is that there is an opportunity for things to vary, for things to change. You can say, there's an opportunity for mutations or for stuff to happen, but another way to look at that is to say 88% of the stuff, in the normal case, is stable. The thesis is that that 12% that is able to change is good: it's healthy, it's directly tied to this idea of competitive advantage and natural selection. It's built into the overall structure to allow us to evolve.

SS: Yes.

HB: It seems to me that you're offering a very different thesis than was previously put forward, which was that we know mutations can occur, and mutations involve basically zapping these little nucleotides somehow, which naturally impinges upon a change in related genes, proteins, and so forth: it's kind of random and can happen or not happen. You're making, it seems to me, quite a different sort of claim, in terms of the structure of evolution.

SS: Well, it's not necessarily new. There are different models of evolution, and they are actually all true in different forms. But it was thought that there were only a handful of specific genes that were under this copy number variable selection, like the amylase genes, for instance.

What we showed is that actually 12% of the genes have this ability to vary in their copy number to contribute to evolution. The change in our understanding is really related to volume. Our research increased our knowledge of the whole set of genes that could benefit, or be selected against, based on their copy number variation.

HB: But it seems to me that there is something philosophically distinct going on here—or maybe I just don't appreciate what the prevailing view was before.

One way to look at it is to say, *"Mutations happen, stuff happens, things are adaptable, you bombard things with different forms of radiation and what have you, and things change around. Who knows what's going to happen. We'll see."*

That seems quite different than another way to look at it, which says, *"There is this built-in structure where we have the possibility for mutations that can be positive—they may not always be positive—but they can allow us this natural evolutionary tendency,"* and then there is the other side of, *"Careful. Don't screw with that stuff over there, because, if you do, bad things will happen."* That seems like a different framework to me.

SS: It was all known before; it's just that we didn't know there was so much copy number variation that could contribute to evolution. I have this figure that I always show the students when I teach.

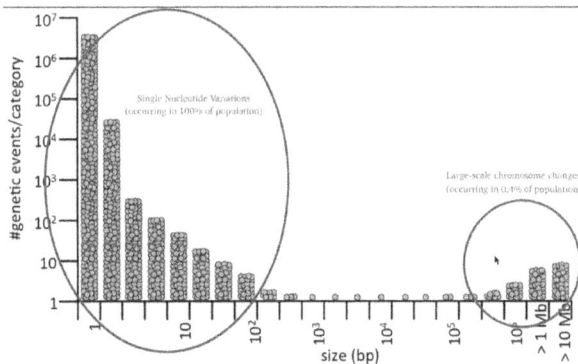

If you plot out the number of different types of genetic variation and divide them into single nucleotide variation and the copy number variations, you'll see that, in fact, 0.4% of the 'normal' population, the average population, carries big chromosome structural changes.

Trisomy 21 is mainly associated with Down syndrome, but there are other big segments of DNA in a very small portion of the population that are different from each other. 0.4% of the population have these big, big changes, and we've known about that for 50 years.

On the one hand we've got those 3.2 million potential single-nucleotide changes we were discussing earlier that everyone is subjected to, and on the other hand we know that 0.4% of the population experience these large-scale chromosome changes.

When I was teaching back in 2002, I kept thinking to myself, *Biology favours balance. There* **must** *be a lot of other variants here. Why is it that we haven't seen them yet?* Well, because we didn't have the tools to see them.

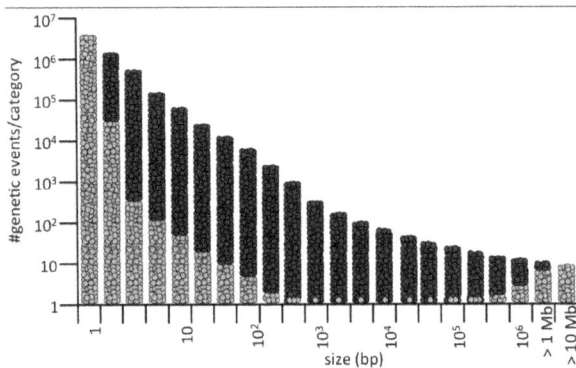

And the sequence comparisons that we got from this new data filled this graph in perfectly.

There were lots of theories around of how single nucleotide variants (SNVs) could affect genes, and there was also an awareness of these big chromosome structural changes that often lead to changes in species, because when you have one of these big changes sometimes you can't actually mate with the same species. But in between there wasn't a clear sense of things.

Now, it turns out that 12% of the genome can be copy number variable. In fact, there are 800,000 different sites along your chromosome, which we didn't know about pre–2004, that can have gene copy-number variations. In some cases, it's the entire gene, from start to finish, encompassing, say, 100,000 nucleotides, and it's present in three copies instead of two.

In other cases, it's an internal segment of the gene where, say, exon number 22 and 23 that encode a motif—a little segment of a protein—have two copies instead of the typical one copy, and that can lead to a slightly different protein.

In some cases, it's a deletion of a segment that encodes a protein that may span a cell membrane 6 times instead of 7 times. It can sometimes be these slight modifications.

Again, we didn't have the ability to see these before. This has really changed the way we look at the effects of genetic variation on both disease and natural evolution.

HB: Whenever I'd hear things like, "*The genomes of different individuals are 99.9% identical*," I'd think to myself, *That's odd.*

It sounded like some sort of counting argument, and I didn't really know what it meant, but it seemed like an odd thing to say when you look around you and you see that even people who are identical twins are so very different. You'd think, *How is that possible?* I don't know the mechanisms, but that seems, just on a very basic level, incongruous. So what you're saying makes much more sense to me.

SS: Yes. It makes more sense.

It's interesting, the first paper we published on the genome-wide copy number variation was 2004, and we've been in the media a lot as a result of our work over the last 15 years or so, but that's the paper that got the least amount of press.

I remember, I did do one interview and thought, *I've to get these guys' attention, because this is our most important stuff*. So when I was riding home the day the paper came out I chuckled to myself, thinking that maybe the big CNV that I carry is what gives me a better golf handicap than my brother, or something like that.

HB: That worked?

SS: That kind of caught on. But nonetheless, we knew that that paper was big.

It was compressed by the journal from a typical 5-figure paper to a 2-figure paper, because they didn't believe it. It took us close to a year to publish that. I remember sitting on my dock at my cottage, on an August day, fighting tooth and nail for hours with the editor of the journal, who had just come back from vacation (he was gone for several months and had forgotten about the paper).

Nobody believed it. Because they all thought, *Why wouldn't we have seen this before?* That's what the reviewers kept saying. "*How could there be that much genetic variation that we didn't know about?*"

But, of course, it's there. These are the sorts of things that play in to our diversity.

We've now looked at monozygotic twins. They were always thought to have the exact same genome—and largely they do—but some of them have gene copy-number variations. These things can occur any time a cell divides, not just when the sperm and egg come together and you get up to, say, the 8-cell stage and tissues start dividing off. These can happen at any time during cell division; and because they tend to be such big changes, they can affect multiple genes at a time.

That's actually what cancer is. Cancer is mainly a genetic disorder that occurs when genes are misregulated—often through these copy number variation changes—in different tissues. Cells get out of control, and then they take over that tissue that they're involved in.

A big area in the field right now is to look at how somatic changes—those changes that occur post germ cell development, later on in organ development or even, perhaps, in brain development—that are specific to that individual, may contribute to a particular aspect of a disorder, or a disease, or makes them special in some way.

We didn't really have the technology to do those studies before, because you obviously can't take brain cells from people, but with brain imaging and so forth, we're getting there.

Questions for Discussion:

1. What do you think Stephen means by, "Biology favours balance"?

2. Why do you think that Stephen's 2004 paper on CNVs got the least amount of press? Is it the fault of the media? The scientific community? Science popularizers? How common do you think it is that papers scientists consider the most significant have the least public impact?

VIII. Causes and Implications

Complexity galore: autism and other conditions

HB: I want to ask a few questions about biological mechanisms. What's our current understanding of how these large-scale variations are actually happening? And then I'd like to look at the mechanistic aspects of what that implies.

SS: We've got some new data coming out based on our whole genome sequencing experiments, both in autism families and in controls. Just to give you some numbers, if you look at any new human being that is born and compare her DNA sequence to the two parents, the baby would have roughly 100 new genetic changes in her genome. That's a rough number. Some of those are copy number variants. Others are single nucleotide changes.

HB: What would the relative percentages of the two be: copy number variants to single nucleotide changes?

SS: It depends on how you define copy number variation. The way we define it, you would have about half a dozen copy number variants affecting roughly 40 or 50 different genes. The single nucleotide changes that would be new to the individual would be of the order of 60 or so per person. That's from the best data so far. And then there is a lot of mixing and matching, but these are the changes we can detect.

Now we can use that information to see which of those are associated with a specific condition.

HB: OK. And what's actually causing those things? What's our best guess as to what's going on?

SS: If you think about the chromosomes coming together during meiosis (that's when you have the exchange of genetic material between the mom and the dad and the pairing of the chromosomes): every time the cell divides, the chromosomes come together. There are 3 billion pieces of chemical information that have to be replicated in the cell, copied, and then pieced back together. It's a very high-fidelity mechanism. Each nucleotide is checked to make sure the right letter is there.

But nature makes mistakes, for whatever reason. If you're exposed to carcinogens, nature makes more mistakes. When you're exposed to x-rays, for example, or asbestos, nature makes more mistakes, because the enzymes that do the checking are compromised and don't work as well.

We've found that, over time, there are segments of chromosomes that are present that look very similar to each other. You can think of the chromosomes coming together in cell division as being sort of like velcro: if there is a segment up here that's very similar in its DNA letter sequence to one that is down there, they can misalign. Then the velcro sort of loops out, and when the enzymes come to replicate the DNA, they skip over these sections and produce a mutation. It's like when you get your zipper caught. If the enzyme is not working so well, because it's been bombarded by x-rays or exposed to carcinogens, you'll have many more mutations. That's how it works.

HB: There certainly seems to be a theoretical possibility to be able to retrodict, to go backwards, and look at masses of people that are subjected to the same environmental effects, or external effects, and map that on to see the specific variation of these mutations. Maybe research is at too preliminary a stage, or maybe it's not.

SS: That's what is happening in studies of cancer—which, again, mainly concerns changes that occur during your life, somatic changes.

But I'll give you the autism story, which I think answers your question.

When we look at families—parents and then children who are autistic and, in some cases, siblings who are not autistic, what we would call a nuclear family or a pedigree—and we run these high-resolution microarrays (we actually get 2.7 million pieces of information across a genome; in fact, now we do the whole genome sequence where we get all 3 billion chemical bases) we find a higher rate of spontaneous copy number mutations in autistic children versus their parents and non-autistic siblings.

But it's not a higher rate across all the chromosomes; it's only a higher rate in the exons of those genes that encode the proteins involved in specific aspects of brain development.

We were asked questions by people who hypothesized that there might be environmental exposures related to autism, for example. We thought a lot about it. It's a really good point. We looked and we didn't see a higher rate across the whole genome. It was only these specific sites, which means that we're seeing it because we're looking for autism and mapping back to those specific changes that link to autism. We did controls; and we see the same general rates across the chromosomes.

Mutation rates are typically random. There are some places where you have what we call hotspots, where these zippers come together, as it were. But typically it's random. Then, when you find a higher rate that is associated with a clinical outcome, the reason is because that's actually the cause, or a risk factor, for that clinical outcome.

HB: You made a comment earlier about these mutations and how we're all very different from each other. In my mind, there are two different points.

The first is that there are these variations, as you said before, within the 'normal' variable stuff, as opposed to the stuff that should be stable.

Then there is the possibility that maybe what seems benign to us is just something we haven't found the negative aspect of yet.

The whole idea of quantifying and qualifying what 'benign' is, seems difficult to do, because you have to be matching it up to some condition, which may or may not be prevalent. It seems like there is a distinction that's being made there. I guess what I'm wondering is, how many things are *really* benign?

SS: I don't think anything is. I think if you have a genetic change that alters an amino acid in a protein, there is opportunity there for it to give you a selective advantage or disadvantage. What that advantage or disadvantage is plays out in evolution.

In most cases—because there are hundreds of thousands of these changes in our genome that affect genes—you don't know. You don't know what makes you look like you, and behave like you, along with your environment; and how those changes interact with your environment. Even identical twins are often so different. It's only when we're looking for an association in a particular instance, usually a clinical outcome or a disease, that we pay attention to these things. There hasn't been a lot of behavioural genetics done for things that aren't associated with a clinical disorder.

The biggest surprise to me in the last ten years has been—and a lot of this came from the copy number variation work—that we're finding that genetics plays a much bigger role in human behaviour than was previously anticipated.

There have been mutations, copy number variants, found in a gene called FOXP235, which is involved in speech and language, for example. We're finding these CNVs to be involved in autism, which is a clinical disorder, but there are lots of people who have Asperger's Syndrome, and variants thereof, who are very high-functioning and incredibly intelligent. Everyone has his own unique features. That's the beauty of genetics. We're all unique. Every 0.1% genetic difference is a lot. That's 3 million nucleotides. That's really the beauty.

What's different about genetics is the fact that, by definition, when you find out information about an individual, you're getting information about the parents and siblings, because that's how genetics works: you inherit these things. That's why it's a different science. It's becoming more quantitative. We're getting much better because our quality of data is better. But to actually translate how that plays out in an organism can be very complex.

HB: These days we talk about the autism spectrum, which is different than how it used to be regarded. From a genetics perspective, I would imagine that would just make things tougher, in the sense of how you define it.

SS: Right. There's a segment of DNA on chromosome 16 that's 680,000 nucleotides long that our group, and a group in Boston and in Chicago, discovered around the same time. In the Canadian cohort, roughly 0.6% of our autism collection would have a deletion of that segment of DNA. There are 17 genes there and, in fact, almost all of them code proteins involved in the brain. That was first observed in autism and intellectual disability. These are people that have one copy instead of two.

Then another study came out a few years later, where they looked at schizophrenia. And they found that, more often than not, people had a duplication of that—so three copies—but also, some had only one copy. Now, either a deletion or a copy number gain—so one copy versus three—has been found in epilepsy, and in bipolar disorder, and in OCD. Most often we see the deletion involved in autism and intellectual disability and the duplication involved in schizophrenia, but not always. It's everything in between. You have lots of other genes that are tweaking how you develop. So it's difficult.

Questions for Discussion:

1. Are you as surprised as Stephen seemed to be at the notion that, "genetics plays a much bigger role in human behaviour than was previously anticipated"?

2. What do you think Howard means, exactly, when he says that invoking the idea of a spectrum would make a condition harder to define?

IX. Towards Treatment

Early diagnoses and equilibrium-restoring medicines

SS: Our primary goal regarding autism is to try to use genetics to identify individuals earlier so that they can be put into early-intervention behavioural protocols, which actually work quite well if you start early enough.

Here in Toronto, there are around 500 kids of about 4–5 years old who are on a wait list to see a proper developmental pediatrician. Well, we've got genetic tests now that could signal something for 20% of those.

HB: And the symptoms often don't manifest themselves until 18 months or something like that, right?

SS: That's right. And if there's a family history, for about, say, 15–20% of them, we can tell them that this particular child of the 500 needs to be prioritized, because he's got this chromosome 16 change, which means that there is a very high probability that he is on the spectrum. So get *that* kid in the system first. They should all be seen, of course, but there are resource issues. That's one of the approaches.

We also naturally want to identify which genes are involved so we can think about which medicines may be used to increase the likelihood of remedying the situation. That's the second approach.

Right now, the approach is to mainly confirm diagnosis and to allow early detection. But, as you said, the challenge is that, even though we can find these changes using these new genetic technologies, it's almost impossible to predict what the outcome is going to be. I use this chromosome 16 change because it has been studied the most, although we only found it a few years ago. There are families that have two autistic children who are siblings: one is extremely high-functioning and one is severely autistic, and they have the same chromosome 16 change. We don't know why. It could be that there are other genes that are modifying the impact. It could have been diet or other things. We're trying to study it.

We're getting more and more data that suggest that the genetic changes and alterations we find tend to lead to more similar characteristics, but even in identical twins who have roughly the same genome, in about 15% or so of cases they, too, can be on opposite ends of the spectrum. There are lots of random effects involved.

HB: Personally, I'm not happy with "random effects". I'm much happier with "not well understood effects", or even "intractable in principle due to complexity" effects. When you biological guys talk about randomness, I typically think to myself that there's some law-like structure out there somewhere but we just don't know what it is.

SS: Well, there's a lot known about how neurons connect, and communicate, and migrate things during brain development; and there is a lot of randomness in it. It depends what hits first, what hits second, how things come together. I guess the way to look at it is that everything has an environment.

HB: So just to clarify, for me, the way I look at what we're talking about in terms of pathogenesis, or "bad stuff happening", or whatever you want to call it, is that this is linked to changes in our DNA that can occur in a number of different ways. It can change at the single nucleotide level or it can change in terms of these copy number variations. And both of these changes can be linked to genes and their consequent proteins, which in turn imply a real physical effect.

But I can also imagine that, when you're trying to assess which one is doing what and where the resulting consequences are there could well be weird non-linearities going on at the protein level.

So you might say, "*We recognize this copy number variation over here and that copy number variation over there, which happen with some frequency and are statistically linked to autism.*"

But then there could be other stuff that we haven't really considered, and those things are interacting at the protein level, and that's what gives you one person with Asperger's and causes a different effect in somebody else. Is that a reasonable way to think about it?

SS: Absolutely. I'll just take that a little further.

When you take all of the genes that make the proteins that we've identified as being involved in autism using this copy number variation approach, for example, you find that there are about 200 different genes that have been identified with different types of copy numbers. Sometimes it's the whole gene, sometimes it's a couple of exons.

Then we use a method called network mapping—because we know a lot about how these proteins interact with each other. Even though there are a couple hundred different genes involved, they're all involved in the same cellular processes.

We've been doing really simple science right now, looking at things one gene at a time. If you have a CNV affecting the "synaptic scaffolding 3 gene", which we know is involved in autism, if you have a deletion in that gene then more often than not (in fact, almost all the time) you'll have some type of an autism spectrum disorder.

But it could be that you have another CNV in a different gene in your genome that may encode a protein that is involved in binding. In fact, there are two other synaptic scaffolding genes, 1 and 2; and there's a whole complex of other proteins that come together. If you have one of those genes that's making more of a protein, for example, it might actually compensate for the two to-one deletion in SHANK3.

That's what we're trying to figure out. We couldn't do that until we had the whole genome sequence. Then it becomes very, very complex.

HB: You need massive computational power, I imagine.

SS: Yes, exactly.

Last summer we published the first whole human genome sequences, and we've got another 1,000 or so that we're looking at now. We estimate that we need 10,000 families with autism, 10,0000 whole-genome sequences, before we can even start to consider how these minor perturbations and other proteins may be modifying the effect of autism at the cellular, and then at the organismal level.

HB: Are there people who specialize in just this, to the extent that they look at the interaction of these proteins: all these potential non-linearities and zillions of possibilities?

SS: Yes. This is what we call omics. We talked about genomics. There are people who do proteomics, who are sequencing all the proteins using mass spectrometry. In our group we've got information scientists, bioinformaticians.

We're getting pretty good at developing computer tools to link the two data sets together. As I said, the problem in studying most disorders is that you can't get access to the tissue that's involved. So you have to do it in animal models and try to link all the data together. There are lots of ways that you can introduce these mutations into cell-based systems and check things. Ultimately, that's exactly what has to be done.

And that's actually the way that new medicines are going to be developed. If we can figure out how to "pump up one of the tires of the car", as it were, it may compensate, for lack of a better term, for "the flat tire on the back left wheel", say.

That's also what got a lot of people excited about the copy number variation story. In most cases these disorders arise due to two-to-one deletions that lead from two copies to one copy, but there are also many variations on that picture.

HB: It gives you so many more degrees of freedom.

SS: Yes. So, coming back to these mechanisms we were talking about earlier, if you can just pump that up a little more you might be able to restore a sense of balance. But in some cases you need to actually decrease it—because the disorder is arising because you actually have too much protein and having three copies is detrimental—so you have to try to bring it down in some way.

The goal is then to understand the whole system and how every thing interacts so that we might be able to provide the balance, or reach equilibrium, in an artificial way.

Questions for Discussion:

1. Might our intuitive notion of "biological balance" be able to be rigorously defined through genetics?

2. What does Howard mean, exactly, by "non-linearities" and how is that related to the need for vastly increased computer power?

X. The Definition of Disease

More complicated than you might think

HB: This may be completely unrelated, but for years I've heard people talking about protein folding. Might establishing this balance more generally have anything to do with protein folding, the geometrical aspect of the protein and how it interacts with other stuff, or is that something completely different?

SS: You're exactly right.

You have the DNA, and RNA is made, and if the RNA doesn't get translated into a protein efficiently—because you have some missing exons, for example—it's degraded in the cell. There are lots of different things that can lead to instabilities with RNA or protein molecules. If they don't become stable, or they misfold, there are whole mechanisms that come into place to get that out of the cell because it's bad.

That has to do with small copy number variance. This is something we really don't understand. I have a new student working on this now: how these small copy number variants are leading to minor changes in the protein structure and function.

HB: How do you define small?

SS: One of the biggest genes in the genome is called the dystrophin gene. It's on the X chromosome. When you have mutations in this gene, it leads to muscular dystrophy. It's mainly a muscle disorder, but causes a number of other challenges as well. But we've been finding that the same gene is involved in a lot of the autistic kids that we look at. It turns out that that gene is very complex: there are 100 or so exons and multiple different forms of the protein that can come together. We're finding lots of these copy number variants that are affecting specific regions of the gene that were never detected in muscular dystrophy before because they're not directly involved in muscular dystrophy. They're involved in an isoform, or a form of the protein, that is actually involved in how the neurons work.

I suspect that there are going to be lots of other variations that lead to other proteins coming together that just don't have a function, and then they're removed from the system. There are many other backup systems in place to get rid of things that are not wanted there.

We don't have a good understanding of all these different combinations. Evolution has tinkered with this for millions of years and we're just looking at the current snapshot. I think what's absolutely fascinating is that our definition of disease, or disorder, or whatever term you want to use, is based on current societal norms.

I gave a talk at a psychiatry meeting a few years ago, and I talked about finding some of these copy number variants in kids with ADHD, bipolar disorder, and schizophrenia. A lot of the genes are the same as we're finding in autism. Some of them are pretty predictive—for example, there is a gene called astrotactin 2 that we see quite often in ADHD.

Anyway, one of the psychiatrists at this meeting put up his hand and said, "*I think I understand. Maybe, because it's so hard to give a diagnosis on these neuropsychiatric or behavioural conditions, we should be running the DNA first to help hone what we should look for.*" And that's exactly right.

The question is, would we still have the same groupings of conditions with the same symptoms had we actually lumped them together in a different way first—by looking at the DNA, say?

The answer is that it would be different. We've actually seen this now. These chromosome 16 deletions I talked about earlier with regards to autism were first described in 2008. Now, when those children are observed in clinical genetics, they're tested using this microarray approach and they then get a different diagnosis: a "chromosome 16 disorder". And they never get into the autism studies. We don't see them in our autism studies anymore.

It's going to be interesting to see how this plays out. At the end of the day, you need to have both approaches.

HB: And the feedback loops go both ways.

SS: They go both ways. And it's just a matter of medical history really, what got there first.

HB: I'm also guessing that, sociologically, it certainly has an impact on both sides. And I imagine that it might well have an impact on how people well outside the medical science community look at things. If they start using this metaphor of a spectrum—together with what you were saying earlier about how everybody has these various mutations and we can categorize them in different ways—then that might lead to a higher level of tolerance on just a humanistic, basic, societal level. Or so I hope, at least.

SS: Yes. And again, that's related to the problem I had with the message from the Human Genome Project. When the draft papers came out in 2000, which talked about how we're 99.9% identical, I remember thinking, *But we're **not** identical. My brother and I share 50% of our DNA from our parents, and we're nothing alike. You could probably pick us out in a crowd, but we're really quite different.*

The CNVs are part of the explanation for the variation. But there are so many things that come into play. I think that it's becoming more and more known that we are quite genetically diverse and there are new changes occurring in our own genomes, even within the cells as they divide as we get older. It's a beautiful thing that we understand that everyone is unique. The genomes are not that similar, actually.

HB: That strikes me as somewhat ironic, because if you talk to somebody from a hard core behavioural perspective, they might well give you the impression that these genetic guys are all out to have a rules-based algorithm that we're all just predestined by our biochemistry and so on.

But it turns out that, when you look deeply at things from a genetic perspective, you recognize the variation and individuality on a societal level, which is not something that you would have expected—at least, it's not something that I expected.

SS: I published a paper in *Nature* in 2008 ("Copy number variations associated with neuropsychiatric conditions") thinking about this concept of how CNVs would be involved in neuropsychiatric disorders like autism. I drew this figure and in it there is this big grey zone. If you have one of these highly penetrant mutations, if you only have one copy of the SHANK3 gene, it doesn't matter if you're male or female, it's usually going to push you across this boundary to make you autistic.

There's another gene I mentioned earlier, astrotactin 2: if you have a deletion and you're a male, often that's sufficient to push you across this threshold and put you on the spectrum. Whereas if you're female, there is some other biology that comes into play—we think maybe hormones—it doesn't matter.

I talked to some physicists and mathematicians and asked them, *"Can we write some formulas, based on big data, and get rid of all that grey?"* It turns out that the best approach is to do these large sequencing and CNV studies and figure out what the likelihood is if you carry one of these alterations, what the probability is that is going to lead to a given condition, so that we can use those statistics in a clinical setting. There are some things—like gender—that are easy: you're male or a female.

But there are all the other things that happen in your lifetime: not only all the "capital E" environmental factors like the toxins you're exposed to, but also the "small e" environmental factors.

I always say the biggest thing that happened in my life was that I was the second born. I spent most of my childhood trying to catch up with my oldest brother, working harder and pushing. That was the biggest event in my life, aside from perhaps getting the Y chromosome and becoming a male. Everything else naturally followed from that. It's funny how you look at it.

But certainly, these copy number variations, and whether you have three copies of a gene or one copy, say, can have a big influence.

Questions for Discussion:

1. Should psychiatrists and psychologists have a thorough grounding in genetics? If so, will the need increase in the future?

2. What do Howard and Stephen mean when they say that "the feedback loops go both ways"?

3. To what extent might we one day be able to distinguish, in principle, between Stephen's "capital E" and "small e" environmental factors?

XI. Probing Deeper

Stem cells, pleiotropy and environmental factors

HB: For people on the outside looking in at what you and your colleagues are doing, I can imagine them saying something like *"These geneticists told us 10 years ago 99.9%. Now it seems they're saying, 'It's a lot less than that; now we're down to 88%.' Maybe 10 years from now they'll tell us it's 60%."*

SS: I think we're very good now at what we call the constitutional genome: the genome that is in each of your cells in your body during early development. But when it comes to the new mutations that are occurring during our lifetime, we don't have a good estimate of that.

There's a really nice paper that came out a couple of months ago that shows that there are a lot more of those types of changes, particularly in those tissues where the cells stay around for a long time, like brain cells. I think that's a new frontier where we're going to see a lot more genetic variation across the whole organism.

The genetic blue print that's the basis for all your cells is similar, but because the genome is so large, there's a lot of variation in it. The beauty is, we have a common ancestry in our DNA of about 99%, but because the genome is so big, every 1% is 30 million nucleotides. That's a lot of variation.

HB: You mentioned these alterations that can occur over our lifetime, and that makes me think about how, analogously, neuroscientists are talking much more about the plasticity of the brain now. Might there possibly be some sort of a link between those two things?

SS: Yes, absolutely. There's a new technology called induced pluri-potent stem cells where you can take a skin cell—that's what we typically use—culture them in a petri dish, then add four proteins, and you can then differentiate them to become pluripotent stem cells that can be transformed into any tissue in the body.

What's interesting is, if you look at what happens when they differentiate them into, say, brain cells versus heart cells or kidney cells, often there are genetic changes that occur that we can detect. I think that's playing into the variation and how each of us is unique in different ways.

There are lots of indications pointing out that this is happening. We haven't had the technology yet that allows us to look at a single cell at a time, but it's actually here now. We're working on this, as are many other groups. I think it's an exciting time. We're never going to know everything, so the important thing is to try to figure out which genetic data is most appropriate to use in a clinical setting. In some cases it probably shouldn't be used because it's not going to be good enough, or it's not going to have enough value for predictability.

But two things surprise me about all of this.

The first is that we're much farther along than I would have thought we'd be had you asked me a long time ago. There are a lot of genes involved in these conditions that we didn't know were even genetic before.

Secondly, what we're finding over and over again is that, for many different clinical syndromes, there are mutations in the same genes. This is a thing called pleiotropy. It could be that the protein is so complex—the protein is involved with spanning the cell membrane and grabbing another molecule, for example, and bringing it in.

If you affect the part that's involved in the cell membrane, it may have a different effect than if you were to affect the part that's grab-bing the molecule and bringing it in. So you would expect a different disorder, or a disorder that is similar but has been described through medical history as a different disease.

In some cases the diseases are actually quite variable, because the gene may be involved in both heart development and brain development. This is really amazing. In a sense, we're back at the stamp collecting stage in medical genetics.

We have to do a lot more of these genotype and phenotype studies, which is essentially what genetics is, and we're much better now at getting the data because we can sequence entire genomes. We understand the concepts of genetic variation and copy number variation now. But ultimately, we have to be able to measure the uniqueness in our phenotypes—our diseases and all the other traits that make us unique. The question is, how far do you want to take it? Is it worth studying something like height, intelligence, skin colour?

In terms of understanding the essence of human nature and what it means to be human, a lot of these studies are really interesting. I do think, too, that many of them will come out of the diseases we study, because these different aspects are involved in disease. They're happening hand in hand. But I think we have to be careful because, at the end of the day, there are just so many factors.

It dawned on me recently that the general public thinks of "the environment" as referring to toxins and other exposures. But environment is everything, as you pointed out earlier.

HB: As you mentioned, the biggest factor for you, was being born second. That's a big environmental factor.

SS: That didn't dawn on me until a few years ago when I heard someone else say it. Actually, it was a song I heard, and it made me think, *That's actually me. That was probably the biggest motivating force in my life, trying to out-do my older brother.*

Questions for Discussion:

1. Might all genes be capable of pleiotropy under the "right" conditions? Can the complexity of the protein be objectively classified?

2. Might plasticity of the brain involve proteins that we currently think have "nothing to do with the brain"?

XII. Ethical and Societal Issues

Towards responsible progress

HB: In terms of medical ethics, are there enhanced concerns that you can see resulting from all of this? Do we have to be a little more careful here; and, if so, how?

SS: I mentioned that there are roughly 6,000 genetic disorders. There have been a lot of studies done over the last few decades on a subset of a few hundred or so where deletions of specific regions lead to very severe clinical syndromes, and there have been lots of ethical and social studies done on those families to see what information they want, what information they don't want, what can be used to help the children, what doesn't work, and so forth.

There is a lot to be learned from the medical literature. Typically you don't learn that in medical school, so that has to happen more. You need to read the literature and learn. I think that's important.

What's changed now—and this is a big change—is that the experiments we do are not really targeted to a specific gene or a specific question, because they are genome-wide. They are hypothesis-free. We get *all* the genetic data—we can sequence a whole genome now for roughly $1,500, and that number will likely come down even further, to $600 or so, in the next few years.

We had a workshop at The Hospital for Sick Children in Toronto to talk about how this data is going to be used both in the postnatal setting and prenatal setting. There are a lot of reasons you would want to know prenatally.

For example, if there is an ultrasound anomaly detected, in many cases the reason for that can be mapped back to a chromosome change—a big copy number variant, for example. Families, for many different reasons, may need to know that. There may be a family history, it may affect the mother's health, together with, of course, the fetus' health.

So you get the DNA—and you can actually do this now by taking maternal blood DNA, because there are fetal cells that float around in the mom's blood. You sequence the mom's blood, put together the genetic code of the fetal cells that are floating in the mom's blood—

HB: Wow. You can actually distinguish between the two?

SS: Yes. The technology is getting better. They are actually starting to use this for copy number variants. You're going to be reading more and more about it in the newspaper, and it's most often done for the right reasons. The technology is improving rapidly.

But when you get the whole genome, you get "The Full Monty": all the information. And what we're finding now is, more often than not, we find a genetic alteration in a gene that's involved in something else than the clinical indication we're looking for. That's even the case in our autism study.

HB: And we're going to find this more and more, of course, as we develop an increased understanding.

SS: Exactly. So, first of all, how do you understand that information? How do you deliver it to a physician who may not really know much about genetics? You get two weeks of genetics in medical school. How do you give that information back to the family who may have anxieties for one reason or another? The technology is really pushing these questions forward.

At this workshop we had a few days ago (the reason I organized the workshop was because of our research on autism) I wanted to know what the people who were working on this prenatal testing were thinking about, because they are now implementing these whole genome tests. The government is paying for it. It's a good thing, as long as we use the information wisely.

The issue that kept coming up over and over again is that they may have ultrasound data, and they might find that there is a structural anomaly, and then typically that would trigger specific genetic testing.

The child may have a chromosome 22 deletion or not, they may have trisomy 21 or not. Now you get all the data. What happens if you find out that the fetus is carrying one of the synaptic scaffolding 3 deletions, or the astrotactin 2, or some of the other genes we've talked about? That's not what they were looking for, but if they get an indication that there is a risk of a genetic alteration that's involved in autism, or schizophrenia, or something that's untreatable, or something that may manifest 50 years down the road, like Parkinson's disease; what do you do with that data?

There is a whole ethical discourse going on all over the world. We talk about these things at every meeting, but it is complicated. Ultimately, the solution is education, but that tends not to get the most investment, unfortunately. I think the technology is going to really be pushing this.

HB: Listening to you, two separate things come to mind.

There is exactly what you've just talked about: becoming aware of all sorts of possibilities that one wasn't necessarily looking for, and what we do with that knowledge.

But then there is this whole question of predisposition or statistical tendency to develop a condition, as opposed to a sense of iron-clad certainty. One imagines that the more we know, the clearer the picture will be that starts to emerge, but that might not be quite the right way to look at it, at least for the near term. Just because our present understanding leads to some greater awareness of a predisposition for something, doesn't mean that that particular thing will actually happen.

SS: And you don't know what that person's environment is going to be, and how all that will play out. All you can do—and, again, this is where the medical literature has a lot of answers—is give the families the information that is deemed most relevant for them. In some cases it may be the whole genome, in some cases it may be more specific. Educate them and help them make the decisions that are most relevant for them. There could be a family history. For example, they may have three kids versus no kids, and that may influence their decision. That's where we are.

The reason I called that meeting together was because things are moving so fast. At the opening of the session I asked, "*Why are you here?*" Because I'm not part of their community; I'm actually the guy who generates the data they use and the technologies they use to make these decisions.

It reminded me of the early days of nuclear physics. I thought, *This is like a Pugwash meeting. We generated this: it can be used for good, it could be used for bad, it could be used for everything. We need to really make sure everybody knows the full power of the data and the technology and that we've got the right people in the room.*

I have to say, it was such a good meeting. People from many different countries came here and gave their advice, sharing their experiences.

Genetics is different than any other science, as I said earlier, because the information is not only specific to the individual, but there are enough of these inherited changes that when you find something in your own genome, it's probably going to be directly relevant to someone in your family as well. Who owns that data?

We work in a pediatric hospital. Parents own that data until the kids are 16. So we have consent forms and things like that. It's a learning process.

The biggest issue right now is actually how you get your genome. I'm a technology guy—or I used to be; I mostly write grants now—and I want to push to get the best genome sequence so we only have to do it once. That is your "genetic report card", for lack of a better term. You've got your report card, your blueprint, but how you define it and annotate it changes every day. So we need to develop good computer algorithms that scan all the medical literature. There are companies doing this, but we're still far, far away.

What are the legal obligations if you give the genome information to the family and then you find out, a couple years later, there is another genetic change that leads to a severe neurological disorder? There might be a drug that would help them, or an insurance company may want to know. There are all kinds of these questions out there.

We've done all kinds of surveys of all the different conditions we study at SickKids and in the adult hospitals. In every case, the families involved in our studies are supportive of what we're doing. They are doing it altruistically. In our autism study, the questions we're asked most often are, *Why is my child autistic? Can you help me explain why this came about in my son or daughter?* Now, in 20% of cases, we can tell them. Before 2004 we couldn't. It's getting better.

The second set of questions we're often asked is, *What is the likelihood it'll come about in my second child, or my sister's child, and how can we prepare for it?* There are lots of different things we may or may not want to do.

And then the third thing, which is wonderfully inspiring, is that so many are so strongly motivated to help other families, and they see the way to help is to give their DNA to contribute to these research studies.

I go to a lot of golf tournaments and that sort of thing for fundraising; and when I go to these things, the parents there will ask me, *"Have you found anything in my DNA?"*

I see them every year. And I say, *"We have 1,600 families. You've got to email me and I'll check."* But more and more now we're giving them answers.

We had a meeting last Christmas and we invited all the families. We had 300. I asked them, *"How many of you, by a show of hands, got an answer from us?"* And there were a lot more hands that went up this year than last year.

But the point is that almost all of them say, *"Just keep working at it."* And then they always say, *"Did you find something that might have helped another family?"* It's really impressive.

We've now connected, via the Internet, some of the families who are completely unrelated. They both have kids who have the same genetic change, the same copy number variant, that came about independently, affecting the same genes in unrelated individuals—in some cases in different parts of the world.

HB: This is public?

SS: No, we don't make it public, but we would email the family saying, *"We found that gene X in individual 342 is the same as your daughter's or your son's. Would you like to talk to each other?"* One family might have a newborn and the other might have a child who is now 18. They might be able to share a lot of experiences, because it's the same genetic change. The family with the older child may be able to help the other a lot by providing information about what worked for them or what didn't work.

This chromosome 16 change I keep referring to—there is actually a whole society of 16p11.2 families who got together on the Internet to share stories and things like that. Melissa Carter, a clinical geneticist at our hospital, has put together another Internet social group for changes we find in a gene called PTCHD1 on the X chromosome, which we discovered in Toronto. It's actually the gene most involved in autism in our collection, more than the 16p change. Melissa has now connected dozens and dozens of families, and they're learning from each other and exchanging stories about drugs or diet and so forth on the Internet, trying to find out what works best.

Questions for Discussion:

1. Should everyone be entitled to know what her genome is?

2. Are you surprised at the notion that "you get two weeks of genetics in medical school"? Do you think that will change in light of how medical science is evolving?

3. Is there a danger that too much emphasis on our genome might lead to a sense of fatalism?

XIII. Future Possibilities

Potentially unravelling biological complexity

HB: You've been very generous with your time, and I'm almost at the end, but I have a few more questions, if you don't mind.

Let's speculate for a moment and go out on a limb here, if you will. Give me a sense of what you imagine might be the case ten years from now—in terms of our growing knowledge and the impact on diagnosis and potential treatment for autism in particular, and then more generally.

SS: I'll start by saying, if you go ten years out we're going to know more but we're not going to know as much as we would have hoped. Even in the simplest systems there is complexity; but in saying that, I think many of the limitations of medicine have been this issue of complexity because we haven't tried to reduce the complexity.

The whole idea of individualized or personalized medicine, which you hear a lot about in the media, is that, for a disorder like autism we know there are hundreds of genes involved. We need to know which form of autism is actually affecting that particular individual if we're going to try to treat it better and in a more specific way.

The same is true for cancers. We're seeing a lot of successes with regard to cancer. Schizophrenia, possibly diabetes, and many other conditions as well, could possibly benefit from this sort of approach.

Now that the technology allows us to sequence the genome, we can have a genetic blueprint, or fingerprint, of what that person's genome looks like, and perhaps be able to subdivide the genetic changes contributing to risk that is related to the disorders that they have.

This has been a real problem in clinical trials. If you're testing drug A on a very heterogeneous population, it's not going to work.

HB: You're waving a really big hammer, basically.

SS: Yes. But if you can subdivide that—and this is what we're trying to do in our autism genetic studies—into a more homogeneous group, it might work better: get through the process better and have more specificity.

HB: Or you'd see if it wouldn't work at all, right? If you have a narrower band then you'll hit more or miss more, presumably.

SS: Right. These studies are really just starting out. The problem was the technology was too expensive before, but the genome sequencing now is really crossing that threshold.

The Human Genome Project and the Celera Project published their sequences in 2000. Craig Venter's was the first personal genome done in 2007. We were a part of that study. It was a great paper. Watson's was done in 2008. And now, I think there are probably over 100,000 genomes that have been done, but they are all over the world in different labs.

The new technology coming out now is going to bring it into the $1,000 realm. This is just going to go crazy. You're going to see this everywhere. We're going to have a million genome sequences in the next couple of years to really allow us to do some pretty good science. If you couple that with the clinical trials, you're going to start to see much more specificity.

Once you get more of these genes teased out, we'll be able to have an increasingly better handle on things like the single gene that can result in multiple clinical outcomes depending on the mutation. We'll have an increasingly larger knowledge base that will allow us to to do much more informed, smarter experiments.

Things will get better. You're going to see new drugs that are being tested for autism based on the gene discoveries. They are still a little while out. Now, what's happening is all the kids who will go into these drug trials will have their genome sequenced too. We're doing that here. We're getting more data. There will be a snowball effect. You'll see many more drugs coming out in the next ten years.

A very good example is cystic fibrosis. We recently marked the 25th anniversary of finding the gene in 1989. 70% of the kids who have CF have the same mutation. It's a three-nucleotide deletion that affects how the protein pumps chloride molecules back and forth across the cell membrane.

HB: 70%?

SS: 70%. I should have told this story earlier. So the mom is carrying the same mutation and the dad is carrying the same mutation. And the reason is, this mutation arose in Caucasians about 5,000 or 6,000 years ago in Northern Europe and provided a selective advantage. So you see this mutation in 1 in 20 Caucasians. You don't see it in Asians, for example, because that mutation didn't arise in that population history.

HB: What was the advantage?

SS: The advantage is thought to be that, during the Industrial Revolution, the carriers were more immune to the effects of pollution, tuberculosis, and things like that. So the carriers actually had a selective advantage.

That gene is at high frequency in the populations that descended from the original mutation and had a positive advantage through the Industrial Revolution. So you don't see CF in Asian populations, and it's very rare in African populations—unless a new mutation just happened to arise, which can happen.

That's why that mutation is so prevalent, but there are actually thousands of other mutations that have accumulated due to these spontaneous mutation mechanisms we talked about—some are CNVs.

About 3-5% of the patients have a specific mutation that affects a different part of the protein, which affects how the protein comes together. It's not very stable, and it doesn't make it to the cell membrane, so it doesn't allow it to pump these chloride channels, which leads to the lung disease.

A drug was developed that actually stabilizes the protein in these kids who have two mutations in both genes and gets it to the cell membrane. It works. For that 3% of patients—I don't want to say they are cured—but they are expected to be healthy and live much longer lives. You can treat it as long as you target the drug to that mutation.

Had you simply thrown that at all the mutations, it would have been diluted out. That's a great example.

Also, that class of drugs can be used in other mutations that look the same in CF, and other disorders of chloride transport, including cancers, actually. There is a protein very similar to the cystic fibrosis protein called a multidrug-resistant protein: if you put too many drugs in it, it actually goes through a copy number change in the cell. You get lots of copies, and then it makes a lot of protein and kicks the drugs out. These proteins are very similar, so you can learn a lot about the different diseases from the clinical trials going on.

I think the key is to somehow develop electronic mechanisms to cull out that important information that goes across disciplines. I really think that we're going to continue to make big discoveries.

The big advances in science are going to be at the interface of these new disciplines enabled by information science.

HB: There is a sociological effect going on too, because you have to have a network which is conducive to flagging various issues so that there is awareness across disciplines so that if you're working over here on something that may turn out to be relevant to somebody over there, they can somehow be aware of it. You are doing so many things over here and you're not sure what might be relevant to them. You're not looking at it from their perspective.

SS: Like everything in biology you need equilibrium, you need balance. What we've tried to do to get this issue of teasing out the complexity of biological systems is to try to have an integrated group. In our genome center that I direct, I have about 100 people. 30% are computer scientists. We also have some physicists in the group.

HB: Physicists?

SS: Yup—looking for patterns: they've got to deal with big data and look for patterns. We have many biochemists and doctors, because they are the ones who actually know the medical questions. Usually the new insights come when there are a couple of us in the room together.

That's one approach. I think you need to try to somehow reduce the complexity into feasible questions that you can pursue. Computers will help you, but I think the human mind is still the best computer. My job is to come up with the ideas, and to figure out ways to try to tease out these complexities, but ultimately there have got to be better ways to do it. It's only a matter of time before the human brain may not be the best way.

HB: Well, we are a ways from that, I suspect. But in the meantime maybe you'll help us get there.

Questions for Discussion:

1. *Do you think that there will always be room for human direction in the problems described in this chapter, or will there come a day when AI systems can successfully proceed virtually independently?*

2. *Is it possible that bold new developments can arise in genetics from the result of a single person or small group, or will it be necessary to have large teams of scientists in order to make significant progress in the years ahead?*

XIV. Contact with Autism

Serendipitous occurrences

HB: You got involved in autism very early before all of this. What was your motivation? How did that happen?

SS: It was serendipity really. When I started my lab, I had worked in this famous cystic fibrosis lab, I worked on the genome project, and I said, *"What am I going to do? I need to get funding."*

It was luck. I was sitting in my office—this was during the days of fax communication. We got a fax from a family from California who had an autistic child who had a translocation. A translocation is one of these big chromosome changes that affect 0.4% of the population that I talked about earlier. He had a translocation, a swapping of a big chunk of chromosome 7 attached to a big chunk of chromosome 13. At the time, that was a great thing in genetic disease discovery, because with muscular dystrophy we used translocations—'we' meaning the community—to find the genes involved in lots of other disorders. So when they faxed me this, I thought, *This is incredible.* They faxed it to me because I was an expert on chromosome 7, as we had discussed earlier.

I remember saying, *"What's this disorder they're talking about?"* I had read about autism before, but I didn't really know what it was. Then, the same day, I looked at the most recent issue of *Human Molecular Genetics*, and noticed a published paper by one of my heroes of science, Tony Monaco.

Tony worked on muscular dystrophy. He was at Oxford, and I had actually interviewed with him: I was going to go there to do a postdoctoral fellowship with him, but then they gave me a job at SickKids so I decided to stay here. Anyway, he published a very nice paper that had used this approach of genetic linkage, which we had talked about with regard to cystic fibrosis, to look for genes involved in autism. And chromosome 7 came up.

The marker that he found was in the same general region of the chromosome as this translocation breakpoint, which is typically the signpost where the gene might be. Back then people thought there was a single gene involved in autism.

And here I was working in a pediatric hospital where there were hundred and hundreds of new diagnoses each month at SickKids, so it all came together. I did a little bit of research. I went to the library, found out what autism was, became fascinated, and we built a program around it.

It turned out that Peter Szatmari, who was at McMaster University, had the best collection of DNA samples of autism in the world, many of which came through SickKids. Peter is now at SickKids. We work together. It was just the right place at the right time. It all came together.

Then we got these philanthropic donations from some key families and we got a few key grants. Now there is an organization called *Autism Speaks* and I'm now directing a 10,000 genome-sequencing project for them.

When the statistics came out about the increased prevalence of children with autism spectrum disorder, the awareness went through the roof. As a result, it's been pretty easy for us to raise funding for our research.

The important thing, of course, is that we're providing answers to the families with autism, but we also test all these new technologies first in autism, because I can pay for those experiments. Remember the early microarray experiments that nearly bankrupted me that I was going to pay for myself if I didn't get the money? Now I have millions of dollars to use for whole genome sequencing.

We were the first in the world to do this. We're always testing the latest technologies on our autism families. That's part of their altruistic contribution to science, and they love it. But we want to provide answers, and that's why we do it.

Questions for Discussion:

1. Stephen mentioned the increased prevalence of autism in contemporary society. Do you think this is because of better testing and diagnosis or a sign of an absolute increase in cases of autism?

2. Are you surprised that a scientist could move so quickly from researching a medical condition in a library to building a top-level research program around it?

XV. Nobel Thoughts

The perks and perils of prizes

HB: I feel compelled now to bring up something a little different because of the timing of our discussion. We're chatting in your living room about your research career, but a few days ago you, Charles Lee and Michael Wigler were put on Thompson Reuter's *ScienceWatch list* as a serious contender to win the Nobel Prize in Physiology or Medicine that will be announced in a few days.

I'm not going to ask you a typically idiotic journalistic-style question like, "*How does that make you feel?*", but something somewhat different.

Are you at all concerned by how, if you do happen to win this thing, your life and your future research career might be changed?

You mentioned a while back, maybe only half facetiously, about how you spend a good chunk of your time writing grants and fundraising at golf tournaments rather than focusing on doing research. Well, winning the Nobel Prize, of course, creates this media firestorm and might make that situation even worse. Does that worry you at all?

SS: It's unlikely it will happen this year. Their full list of so-called "Citation Laureates" in our field has maybe 50 or 60 people on it. It's based on the impact and the number of times your papers have been cited. It's quite a good predictor for winning big prizes in the long term, but the short term is much less certain.

Let me put it this way: Monday morning I have a duct cleaner coming. My wife asked, "*Should we cancel?*" and I said, "*Nah, I don't think so.*" And if it did turn out to happen, I'm sure they would understand and go away.

If you look at the other people on this list—like McCulloch and Till, the two guys who discovered stem cells, or my good friend Tony Pawson, who sadly passed away last year—these are giants.

But if you do win, I think it becomes an enormous responsibility, and I think the key is that you want to be prepared for it.

In the last ten years, I've been in a very fortunate situation where I have a lot of funding to do my research and I've always tried to promote studies of ethical, legal, and social issues and their implications.

My whole career I've been a *"Just Do It"* guy—always pushing the technology and striding forward quickly. But then you get a little bit older and you think about the cases where there are experiments that *shouldn't* be done. Should I be the person making those decisions or should it be a committee? In some cases it has to be an individual standing up to say, *"This is what should be done; this is what shouldn't be done."*

That's why I think that these kinds of prizes are so important. You get the respect, and notoriety, and you get the ear of the right people. I think it's critical that if you are lucky enough to win one of them, you treat it as a responsibility. But I haven't had time to think about that specifically.

HB: You won't have time if you win, I suspect. You'll probably have even less time.

SS: That's probably true. But I guess what I'm trying to say is that, because I was exposed very early on at every stage of my career to being around these kind of special people, I saw them lead by example.

As I said earlier, a big event that happened very early on in my career was my experience in the cystic fibrosis discovery. That was the biggest race, even arguably bigger than the Human Genome Project—it was a huge thing in the 1980s. I had the privilege of having a front row seat to watch people like Francis Collins and Lap-Chee Tsui and others. I got to see how they negotiated things like authorship. I learned a lot very early.

I've been very lucky; and increasingly I think that my role is to try to get this message out to my trainees so society can understand what goes on in the minds of scientists. As I said earlier, I think that scientists need to be much more broadly educated, and the key is not so much to impose things like research ethics boards: it's to make sure they have the right common sense so they make the right decisions at least most of the time—you're never going to get it 100% right.

If it turns out that I get more power to positively effect things because I have a medal around my neck, I think that's a good thing. The pressure that comes along with being a good advisor, speaker, and advocate—well, we'll see how that plays out.

I got this cryptic email from Thomson Reuters three weeks ago. Of course I had heard about the list—we all watch it each year. My friend Tony Pawson, who was picked for it a few years back once said to me, *"Getting on that list is the second most important thing that can happen. If you wind up winning the Nobel Prize, that's obviously the first. But just being on that list is a tremendous accomplishment."*

There was a press release around the publication of the list last Thursday, and it's just consumed much of our lives since then. In a way, I'm hoping it doesn't happen for a long time. Of course, I hope it happens eventually, but maybe ten years out so I can live out another ten years of my scientific career without these distractions. We'll see.

Questions for Discussion:

1. Are there any negative aspects associated with the Nobel Prizes?

2. What do you think Stephen means, exactly, when he talks about the need for scientists to be "more broadly educated"?

XVI. The Human Condition

Genetic windows on humanity

HB: One last question, which is a favourite of mine to ask. If I were an omniscient being and could give you answers to two or three fundamental questions, what would those questions be?

SS: I think this issue of nature vs. nurture is what all of us are interested in. In some cases we study different aspects of how DNA and proteins contribute, while sociologists and historians study different aspects of the events in one's lifetime or in history: *what has the biggest effect? Is cultural change contributing more than genetic change or evolutionary change?* I'm starting to think about these things. I'm at the stage of my career where I can afford to do that.

A real challenge will be to think of new ways to capture data to look at this. I think the Internet is great—you look at how the number of website hits tracks with disease, for example, or more behavioural aspects or even how we order things online—but the trick is to really couple that to the big questions like nature vs. nurture.

Is it going to be possible? Should we even do it? I think we should. I think this is what makes us human. Will we be able to design a computer that can out-compete the brain? Will it even be based upon what we now know about the brain?

How far should we take the genetics? There is a technology that has come out in the last two years that is unbelievable. We talked about enzymes and sticking DNA together earlier—there's a technology now that allows you to make essentially any mutation you want in a cell—not only in the petri plate, but actually, perhaps, in vivo in a human or a mouse, or whatever. It's called RNA editing technology. It will win the Nobel Prize very soon.

It's come on the stage like a storm, and it's actually very easy to do. I have a postdoctoral fellow who got a big Banting Fellowship to do this. When I asked him how it was going, he told me, "*It's easy!*"

We've taken mutations that we've found from cystic fibrosis patients, and now he's got this working in weeks. We can introduce these in the lab. It took years before. To be honest, you couldn't even do it before with that specificity.

That's something that came right out of the blue. I could not believe we could do this. Now I think, *What's the **next** big thing?* I don't know, but these discoveries keep coming.

I had Craig Venter come and give a talk at SickKids a few years ago and I asked him at the end, a very similar question to what you asked me. There were a thousand people in the audience, so he wasn't too happy with me. The title of his talk was *Reading and Writing the Genome*. We can read it now, but what about writing it? Should we take genetics forward and use it to enhance the human species? You can define 'enhance' in a lot of different ways. It could be athletic, it could be intellectual, it could be something in between.

I knew he had thought about these things a good deal. He told me, "*That was just private talk over a beer. You weren't supposed to ask me that in front of all these people!*" I said, "*Come on, we don't get to see you very often. We want to hear what you think.*" We ended up talking quite a bit about it and the audience got quite involved.

I think it's fascinating: we now have the tools and the information to alter human evolution. We might get it right, we might get it wrong, but evolution is ultimately the game that makes the decisions for us.

I think the key question is, *are you going to need to use all of this to save the species?* There has been a lot in the media about this in the last few years. There are a lot of big thinkers who think that perhaps we're not going to be around a hundred years from now for environmental reasons, or war, or whatever it might be. Perhaps it's time to start thinking about those things. I don't think we know enough yet, but some of these technologies are really moving things quickly. Genome sequencing, too, is raising lots of new questions.

At the end of the day, I think common sense should prevail, and I think we should educate the scientists in the same way that I think humanitarian and religious leaders were very broadly educated in the past. I don't think most scientists get that.

When I was a student at the University of Waterloo, I learned a lot of technical stuff. I didn't learn a lot about history and sociology. I think we need to train scientists as "professional scientists"—not so much as a physicist or molecular biologist, or whatever—because people are looking to scientists now for the answers. And I think that's key.

HB: Anything else? Anything we missed or you'd like to add?

SS: Well, to come back to one of the big questions: one of the reasons I'm so interested in autism, of course, is to try to help the families. I don't have any experience in my immediate family history, but it's incredible how many people have had the experience of someone in their family having autism.

I think it's such a fascinating condition to study, because it encompasses the core features that make us human: language (40% of autistic kids don't speak), cognition (many autistic individuals have severe intellectual disability) and social interaction skills, the ability to communicate with each other, which is arguably what has made us such a successful species.

We see these kids with autism come into the clinics and we try to help the families with answers, diagnostics, treatments and perhaps medicine. But the information we're revealing really maps back to the core questions of what make us human.

This is fascinating for me. I'm a geneticist, which is essentially a historian of our DNA. To get a glimpse into the history and the sociological history of the human species through the disorders we study, is really fascinating. The genes we're finding, the proteins they make and what they do in other species, that's all going to inform our understanding of autism—but more generally it will inform our understanding of what it means to be human. So I couldn't be in a better career right now.

HB: That's great. You've summed that up beautifully. Thank you very much, Stephen. I really enjoyed that.

SS: Thank you, Howard.

Questions for Discussion:

1. Has this conversation made you look at "disease" and "medical conditions" in a different way?

2. By training young people as "professional scientists" as Stephen suggests, do we risk having them not learning as much about their scientific specialties as we need to make progress?

*3. Should humanities students and those trying to deepen their understanding of "the human condition" study genetics? Readers interested in this question are referred to the Ideas Roadshow conversation, **Embracing Complexity**, with Princeton University historian David Cannadine (Chapter VI).*

Sleep Insights

A conversation with Matthew Walker

Introduction

Dreams of a Final Theory

Why do we sleep?

It's such a widespread, everyday phenomenon that it's natural to take it for granted. But when you stop for a moment and think about it, puzzles abound.

Matt Walker, Founder and Director of UC Berkeley's Center for Human Sleep Science, puts it this way.

> *"When you just back up and think about sleep generally, it seems like the very worst thing that could have ever been designed. You're vulnerable to prey, you're not finding food, you're not socially interacting as a group, you're not finding a mate. It looks disastrous.*

> *"There's a great quote from one of the founding fathers of sleep, Allan Rechtschaffen, who said, 'If sleep doesn't serve an absolutely vital function, it's the biggest mistake evolution ever made.'"*

So is it a mistake?

Not very likely.

For one thing, as Matt is quick to point out, sleep has been observed in every species that we've studied to date, which looks like overwhelming evidence of its universal significance: if sleep was not evolutionarily advantageous, surely other fitter, non-sleeping species would have emerged somewhere to conquer their sleeping counterparts, which presumably would have been pretty easy, given that their competitors regularly lie immobile for long stretches of time. But none have.

Which means, according to Matt, that far from being an evolutionary mistake, sleep is nothing less than a fundamental, core biological drive, together with eating, drinking and reproducing.

But still, what does it **do**, exactly?

Well, we don't have a complete picture, but recent strides in our understanding have been enormous.

> *"I would say that twenty years ago the question was, What is the function of sleep? Nowadays, that question is essentially being turned on its head, Is there anything that sleep does **not** benefit? Is there anything that **doesn't** go awry when you don't get sleep? And is there anything that **doesn't** gain an advantage when you do?"*

Matt's lab at Berkeley has been focused on the impact of sleep on brain function, but he is quick to point out that other core processes of the body also benefit from the restorative process of sleep.

As far as the brain goes, current research indicates that sleep is essential to two core functions: learning and memory, and regulating emotions.

When it comes to learning and memory, it seems that sleep acts like a sort of vital "cleansing agent", allowing us to transfer recently acquired memories to long-term storage while refreshing our brains to enable new learning and memories during the upcoming day.

The metaphor Matt uses to describe this process is that of a USB stick transferring data to a hard drive. New memories and learned processes that are acquired throughout the day are lodged in the hippocampus region of our brain, which has a limited storage capacity and thus acts like this metaphorical USB stick. During sleep, this information is somehow transferred to the vastly larger storage space of the cerebral cortex, thereby simultaneously saving memories and thoughts for the longer term while also freeing up space in the hippocampus for the next day's experiences.

But, even more intriguingly, that's not all.

"What also seems to be happening during sleep is that you're testing out associations. It's a little bit like memory pinball. You're taking what you've learned during the day and you're launching it up into this attic of everything you've learned in your past experience—the cerebral cortex—and you bounce it around these nodes up there."

It is this sort of activity, Matt speculates, that every so often allows us to come up with interesting insights while we are sleeping. After all, he pointed out playfully, *"When was the last time anyone ever told you that you should stay awake on a problem?"*

In summary, then, there seems little doubt that sleep has a profound effect on the brain's capacity for both learning and memory. But what about this second aspect of "emotional regulation"? What's that about?

"The first thing to appreciate is that emotion, unlike learning and memory, involves a much greater amount of fine tuning. With learning and memory, presumably the more you have, the better. With emotion, on the other hand, both too much and too little are no good. And we're starting to realize now that sleep will enable you to be in the optimal emotional state between too little and too much."

Extensive neuroimaging experiments with subjects entering the scanner with varying levels of sleep the previous night revealed that showing a series of unpleasant images to those who were sleep-deprived provoked a significantly stronger response in their amygdala—the brain region strongly associated with emotional reactions—compared to their well-rested colleagues.

The current thinking is that these amygdala responses are actively regulated by a very different area of the brain in the frontal lobes, with sleep somehow playing a vital role in the maintenance of the active connection between the two areas thereby ensuring that your knee-jerk emotional responses are put in proper perspective and don't disproportionately govern your state of mind.

But how does sleep actually do this?

The details are still fuzzy, but Matt believes that somehow sleep enables us to reprocess emotional experiences without the associated stress chemistry, thereby allowing the brain to rationally evaluate our daily emotional experiences and place them in their proper context.

If that sounds both vague and complicated, it's probably because it is. Suffice it to say that we've got a long way to go before a complete understanding of the function and specific mechanics of sleep are fully understood. But one thing is clear as glass: however sleep works in detail, we simply couldn't function without it.

Sleep as an evolutionary mistake? Wake up and smell the coffee.

The Conversation

I. Awakenings

From dementia to sleep, and now back again

HB: How does someone become the director of a sleep lab? How does your career trajectory take you to that? Had you always dreamt—as it were—of going into sleep research?

MW: That's a good question. I don't think anyone ever wakes up when they're five or six-years-old and says, "*I want to be a sleep researcher.*" It's not quite the same as saying, "*I want to be a doctor. I want to be a fireman. I want to be a nurse.*" I think everyone in my field is an accidental sleep researcher: they came to it through some sort of circuitous route.

For me, my PhD was actually concerned with investigating dementia and trying to differentially diagnose people with different types of dementia using brainwave patterns. We would apply electrodes and look at the brain oscillations to see if, very early in the course of those disorders, we could separate out which form of dementia they had so that they could be treated effectively.

During that time period, I felt that I was failing miserably. I had no results, no data, nothing looked particularly useful. So I started to study the pathologies of these different types of dementias. Some of them turned out to actually hit centres in the brain that would control sleep, whereas others would be elsewhere in the brain.

I started to think that maybe I was just measuring brainwave activity at the wrong time: I was measuring it while these people were awake. Instead, maybe I should be measuring it during sleep. I applied for a small grant to investigate further and was then able to set up a sleep lab by myself. I really didn't know what I was doing. It was sort of just trial and error.

HB: There were no other sleep labs that you could take as models?

MW: In the UK at the time—this was roughly 20 years ago—sleep research just hadn't taken hold. It's better now, but at the time there were very few sleep labs in the country writ large. So you really had to just fumble through and learn it yourself.

HB: So you became an inadvertent pioneer.

MW: I wouldn't say "pioneer". But I was trying to be a jack of some trades, at least.

Then we started to see demonstrable differences in terms of patterns of sleep brainwave activity. That was great, but that then led me to the next question: *If sleep is problematic in some patients, then what is the function of sleep?*

If they aren't getting sufficient amounts of sleep, and sleep is serving some brain functions, maybe that's one of the contributing factors to their dementia. Then I started to read a lot about sleep and I realized that, in fact, nobody could give me a satisfying answer to why we slept, which just blew me away. It was remarkable. At that point I thought, *This is a perfect postdoc. I will go and study with someone for two years and understand how it works.*

That's how naive I was: I thought that within two years I would solve a question that many very smart people had tried to answer for their entire careers.

HB: That's the benefit of youth: you don't know that you don't know.

MW: My ambition far outweighed my talent—and still does. Anyway, that was how I began trying to understand sleep and its basic functions. That was 14 years ago now and I'm still trying to answer the question of why we sleep.

When I finally get there, I think I'll be able to come full circle back to the dementia question. But right now I'm still trying to figure out sleep.

So to answer your question, I became a sleep researcher completely by accident, navigated by question marks.

HB: But are you still looking specifically at some pathologies and their impact on sleep?

MW: We are.

HB: We'll get to that a little later on, I hope. But it's perhaps worth mentioning that it's not as though you've completely forgotten about those other aspects.

MW: We're actually now finally getting there: we're coming back to the dementia question. It took me a long time, but we're getting there.

Questions for Discussion:

1. Are you surprised at the idea that sleep research was not very well-developed until relatively recently? What other areas of science might we say that about twenty years from now?

2. Might "accidental researchers" be generally more passionate about their work than those who held a long-idealized view of what they would become?

II. Stages of Sleep

Deconstructing sleep architecture

HB: Let's talk about what we actually ***do*** know about sleep, both in terms of sleep patterns and corresponding brain processing. I've heard the words REM sleep, non-REM, and so forth being bandied around. Maybe you can start off by telling me what these things are and giving me a sense of the different stages of sleep, or at least our current understanding of all that.

MW: Sleep, at least within mammalian species, has been broadly separated into two main types. On the one hand there is non rapid eye movement sleep, or NREM sleep. NREM sleep has been further subdivided into four separate stages, which have unimaginatively been called stages 1 through 4, which increase in depth. Stages 3 and 4 are those really deep restorative stages of sleep.

HB: How do you quantify depth of sleep?

MW: Good question. That word "depth" really has several angles of explanation. One of them is that your sensory threshold for being woken up is actually raised, so as a consequence it's harder to bring you out of stages 3 and 4 sleep, and easier to bring you out of, for example, stage 1 NREM sleep. Stage 1 is that light sleep onset phase where even just the creak of a door brings you back up, whereas when you're in stages 3 and 4—deep sleep—the creak of a door is very unlikely to wake you up; you're deep down in the depths of sleep.

The other reason that we call it deep sleep is because we see these big, slow, lazy brainwaves that are happening during this stage of sleep. When you're awake, your brain is going up and down, in terms of its electrical activity, maybe 40, 50, 60, or even 70 times per second, whereas when you go into deep sleep, slow-wave sleep, stages 3 and 4 of NREM, your brain's electrical activity only goes up and down perhaps 3 or 4 times per second. It's really slowed down. It's not dissimilar to some patterns of coma.

The problem, by the way, is that this led us to believe that deep sleep, slow-wave sleep, stages 3 and 4, was just a time when your brain was dormant and not really doing anything. In fact, nothing could be further from the truth. It turns out that what's happening during this time is that massive areas of your cortex, this folded mass of tissue on top of the brain, all decide to chant and sing together in unison. It's a highly synchronized pattern that never happens when we're awake.

An analogy would be if you were at a football stadium and there was a single microphone dangling over the middle of the stadium, which represents the electrodes on top of the head. Underneath you're going to be measuring the summed activity of hundreds of thousands of people. Before the game starts, each of those individual people in the stadium are talking to each other, having different conversations, processing information, very much like when your brain is awake—the chatter is frenetic and fast.

When you go into deep, slow-wave sleep, though, it's as if the entire stadium, all of those people, have decided to start chanting together in a very slow, rhythmic manner—if you were at Berkeley, for example, that chant would be *"Stanford Sucks."*

But what's remarkable is that they've all been synchronized: they've all stopped their individual chatter and fast-frequency activity, and they're all simply engaged in this slow, rhythmic, almost meditative, synchronized activity. We're only now starting to understand exactly what that synchronous activity is.

So to return to your question, when we talk about deep sleep that is characterized both by what is required for you to wake you up and also in terms of these slow, synchronized brainwaves.

HB: This brainwave activity is across the entire brain?

MW: Yes, across the brain. They principally start in the front of your brain, in the frontal lobe, and then seem to regress back. They can start anywhere, but the majority of these deep, slow-wave sleep waves are over the frontal cortex. It's a very strange state.

However, compare that to rapid eye movement sleep, or REM sleep. In terms of brain activity, REM sleep is indistinguishable from your brain activity when you're awake. In fact, some parts of your brain are up to 30% *more* active when you're in REM sleep compared to when you're awake.

HB: Are those the same parts of the brain that you were talking about before, the ones that are related to dementia?

MW: Not necessarily. Those are deep, emotional centres of the brain; and that has started to point us towards some of the functional reasons why we have REM sleep, namely its role in emotional processing, which I hope to talk about a little later.

REM sleep is sort of a contrasting state. It's what we would often describe 30 or 40 years ago as "paradoxical sleep"—paradoxical, because you are recumbent and seemingly non-conscious, but your brain is frenetic in terms of its activity.

One of the other interesting features of REM sleep is that the mechanisms that control REM and NREM sleep are deep in the brain stem, which will send signals up to your cortex to express these patterns of either deep, slow-wave sleep, or active REM sleep.

But when you're in REM sleep, there's another signal that's sent down into your spinal cord, and that signal paralyses all of the alpha-motor neurons in your spinal cord. In other words, what that results in is utter paralysis of your body. All of your voluntary skeletal muscles are inhibited during REM sleep. Now why would your brain do that?

HB: That's my question. I wanted to ask you about evolutionary arguments, but this certainly seems like a very counter-evolutionary situation.

MW: Yes, it's a very strange thing. However, REM sleep is the principal stage during which you dream, so it's a safety mechanism. What you don't want to be doing when it's dark, your eyes are closed and your ears are not perceiving the outside world, is acting out your dreams in your environment. So, very cleverly, your brain paralyses your body so your mind can dream safely. You stay in one place.

These two types of sleep, NREM sleep and REM sleep, will then battle for brain domination throughout the night. That sort of cerebral war is going to be won and lost every 90 minutes in humans, and then replayed every 90 minutes. What that creates, essentially, is what we call a sleep cycle, and across the night you get a standard architecture of sleep.

HB: So this is the "sleep architecture" that I've heard people talk about?

MW: That's right. So you don't simply have all of your deep, NREM sleep first for five or six hours, and then all of your REM sleep second, or vice versa. They keep flipping back and forth. This "war", as I described it, just keeps going on and on: won and lost, won and lost.

HB: How much variation is there from person to person? Is it roughly constant for everyone, or do some people have very different cycles?

MW: It's quite variable from one person to the next. However, within any one individual, it's remarkably stable from one night to the next. In fact, some people have argued that the electrical signature of your sleep—not just the stages and the architecture, but the electrical map, if you will, of brain activity during sleep—is so highly replicable from one night to the next, that it's almost like a sleep fingerprint. It's that specific and that repeatable.

HB: Even if people take drugs?

MW: Well, once you start playing with pharmacology, as well as things like alcohol, then things will change considerably, which we can discuss later. But, for the most part, as long as you're "clean", as it were, that electrical signature is consistent within one individual from one night to the next.

It does change across the lifespan though. That's one of the interesting features of sleep. How you slept when you were younger is not the same as how you sleep as an adult, and it's not how you will sleep when you're older.

When you were young, for example, the timing of your sleep was different: you would go to bed early and you would wake up early. Then, as you go through adolescence, that shifts forward, so you want to go to bed late and wake up late, despite school times not allowing you to do that. That's a whole issue in and of itself. Then, as you get older, that starts to shift back again, so you start to want to go to bed a little bit earlier and wake up a little bit earlier.

However, there are remarkable differences from one individual to the next, even within the same age range. Some people are what we call "owls": they like to go to bed late and wake up late. Others are "larks": they like to go to bed early and wake up early. And we know the basis of that is genetic.

HB: Looking at the "sleep fingerprint" of a particular individual as they get older, is it as if the cycle of the REM and NREM sleep just gets compressed in some way? If you were to actually look at that architecture, is it roughly the same as we age, or does it change drastically?

MW: It changes quite drastically. Within the space of, let's say, six months or a year, there's a good degree of stability, but once you start to fast forward in time three or four years, your sleep starts to change in significant ways, from early adulthood into your 40s, 50s, and so on.

HB: By looking at someone's current sleep fingerprint, can you predict how their sleep architecture will change as they age? Does it change in a predictable way?

MW: That's a great question. It's a fundamental question actually, and the answer is that we simply don't know.

It's an interesting question, because it suggests that sleep early in life may be an interesting bio-marker of itself: how it will be resilient, or vulnerable, to deterioration over time, and how vulnerable the other functions that we now know sleep supports will be as a consequence.

I think it's a very important question and one that we don't yet have a very good understanding of. Is there something about sleep as a seed early on in life that then germinates into a predictive biological factor which determines changes in sleep itself and subsequent disease states related to sleep deterioration? We don't know the answer to that.

What we *do* know is that perhaps the most remarkable changes in your sleep as a result of age are in the domain of deep, slow-wave sleep—stages 3 and 4. You can already start to see that the quality of that sleep—in terms of how big those brainwaves are and how much of the night that slow-wave sleep comprises—is starting to change in your early 30s. That's how early these changes start to occur.

By the time you're 50, you will have lost almost 50% of that deep sleep that you would have had in your early 20s, on average. By the age of 75 or 80, you only have about 5–10% of your original deep, slow-wave, stage 3 and 4 NREM sleep existing anymore.

We've known about that deterioration for many years now, we've been able to plot that. But now, because we're starting to understand exactly what the functions of sleep are, particularly the functions of these different stages of sleep—different stages of sleep have different functions—we're starting to understand that such a change in sleep comes with a cost. It's not as though our sleep changes and everything else remains the same. Functional abnormalities occur because of that, and we're just now starting to unpack those.

HB: I have a couple of questions based on what you've just said. As you were talking about genetic markers, an obvious thing to look at is twin data. Have there been any studies that have specifically looked at the particular individual sleep architectures or sleep signatures of twins? Will I find that twins tend to have similar sleep patterns, or is that not actually the case?

MW: That is actually the case. People have even closely examined individual electrical bursts of activity that are unique to sleep, little inflections of electrical activity.

For example, there's something called a sleep spindle. This is a short burst of synchronous activity that lasts for about a second. Your brain waves will be going along in NREM sleep—the lighter stages, normally stage 2—and all of a sudden your brain will have a little burst of activity called a sleep spindle.

We've known about it for over 50 years and we're starting to understand that it, too, has a functional purpose. But the point here is that even these little bursts of activity—forget even the gross architecture of sleep, which is highly stable within monozygotic twins— even their electrical bursts of brain activity within sleep itself, are remarkably similar based on that homology of genetics. So there certainly is a strong, heritable, genetic basis for our electrical sleep patterns.

Questions for Discussion:

1. Have you noticed your sleep patterns changing? If so, how would you describe this change using the terminology introduced in this chapter?

2. Why do you think that Howard was particularly interested in sleep studies on twins and why does Matt specifically mention "monozygotic" twins in his response? How might such studies help us better establish the impact of personal experiences on our quality of sleep?

III. Parasomnias and Evolution

Getting it right, most of the time

HB: When you were talking about this response mechanism that the body has to anaesthetize you, or temporarily paralyse you, when you're dreaming, I was thinking, *What about sleepwalkers?* Are these people lacking something? Can you point to some particular congenital issue with regard to them?

MW: Very good question. One of the misconceptions about sleep is that sleepwalking and sleep-talking are happening during REM sleep—when they're dreaming. It's not. Instead, what's happening is that these conditions are coming from the deeper stages of NREM sleep, stages 3 and 4.

What seems to be happening is that your brain is in this deep stage of slow-wave sleep and then something seems to try and trigger you to wake up: either an internal stimulation—the nervous system has a burst of electrical activity like a lighting bolt—or maybe an outside stimulus. The brain then tries to go from the basement to the penthouse of wakefulness but gets stuck somewhere between. Meanwhile, the brainwaves, as you're measuring them in the lab, still look nice and slow and lazy. It would suggest that the patient is in deep, NREM sleep, yet you can see on the video camera that they're walking around. They look like they're exhibiting waking behaviour. So sleepwalking, sleep-talking, sleep-eating—

HB: Sleep-*eating*?

MW: Absolutely. There's even sleep-texting now. You can see that the brain is still in deep, slow-wave sleep, yet these people are enacting what seem to be very routine, rote behaviours. It's nothing complex. They'll go over to the refrigerator, open the door, get a glass of water, put it to their lips, and put it down. They'll walk around the room. These are what we call *parasomnias*, in other words, "disorders around sleep."

If you wake someone up who is having a sleepwalking episode and ask her to tell you what was going through her mind, she often cannot tell you anything. Nothing was in her mind at the time. In other words, she wasn't dreaming; she was just in this sort of auto-pilot-mode of motor behaviour.

That's sleepwalking and sleep-talking. However, that's not to say odd things can't happen during REM sleep. They can and they do. There is another disorder called REM sleep behaviour disorder. In that case the paralysis which locks the body into place—at least all of the voluntary skeletal muscles—starts to become impaired so these people, as they are going into REM sleep, start to act out their dreams.

The reason seems to be that the brain-stem mechanism, which releases a signal down into the spinal cord to inhibit all of those muscles, that part of the brain starts to deteriorate. And that degrad-ation prevents that signal from being sufficiently triggered, so these people start to act out their dreams; and they can be quite violent. People have injured their spouses. There are even remarkable cases where people have murdered their spouses in their sleep. It's unthinkable.

HB: So this is obviously a reason why, for most people, that *doesn't* happen. This seems clear evidence of how evolution has protected us by producing mechanisms to paralyse people when they are dreaming.

MW: Absolutely. It makes complete sense.

If you back up and just think about sleep generally, it seems at first glance like the very *worst* thing that evolution could have ever designed. You're vulnerable to prey. You're not finding food. You're not socially interacting as a group. You're not finding a mate. It's disastrous.

There's a great quote from one of the founding fathers of sleep, Allan Rechtschaffen, who said, *"If sleep doesn't serve an absolutely vital function, it's the biggest mistake evolution ever made."* All of these remarkably complex mechanisms must have evolved to keep us in sleep—deep, slow-wave sleep, REM sleep, paralysis, and so forth.

That involves a vast evolutionary cost. What that tells us is that sleep, to have persisted through evolution, to have fought its way through despite all of those negatives, must be essential at the most basic of biological levels.

In fact, another argument in its favour is that sleep has been observed in every species that we've studied to date. It's pervasive. Again, if it is so common across phylogeny, it must serve some very important basic function.

Sleep, together with eating, drinking, and reproducing, is the fourth main biological drive. What's striking is that we've known the functions of those other three for hundreds, if not thousands, of years. But the reason that we sleep—the fourth biological drive—remains a mystery.

Questions for Discussion:

1. What, exactly, does Matt mean when he speaks of "a vast evolutionary cost"?

2. Does Allan Rechtschaffen's argument serve as a "proof" of evolution or a proof of the importance of sleep assuming evolution is correct? Or, somehow, both?

IV. Learning and Memory

Three vital aspects

HB: This strikes me as a perfect lead-in to some of your own research. Let's look at your investigations into the functions of sleep in terms of the link to memory, emotions, learned acquisition of skills and so forth. Maybe you could start by talking a little bit about what you've learned and what the current state of research is in terms of the concrete evidence we have for believing in some specific aspects of the utility of sleep.

MW: Well, 20 years ago, the way the question was posed was, "*What is the function of sleep? What does sleep do?*" Now, 20 years later, the question has essentially been turned on its head: "*Is there anything that sleep does **not** benefit? Is there anything that **doesn't** go awry when you don't get sleep? And is there anything that **doesn't** gain an advantage when you **do** get sleep—seemingly every tissue in the body and every process in the brain?*"

What we've been focused on is the benefit of sleep for brain function. There seem to be at least two benefits. One is in information processing, learning and memory, and brain plasticity; and the second is in regulating our emotions—preparing us for the next day's social and psychological interactions, which have a very strong thematic connection to psychiatric disorders and sleep disruption.

Perhaps I'll first address sleep and memory. What we've understood to date is that sleep is important for at least three different aspects of learning and memory.

The first is that you need to sleep before learning to prepare very specific parts of your brain so they're ready to soak up new information the next day, essentially like a dry sponge.

In other words, people who think they can pull an all-nighter and then learn effectively are desperately delusional. If you take sleep away, just for one night, the ability of your brain to learn basic facts, textbook-like facts, is decreased by 40%. If you want to put it in an educational context, that would simply be the difference between acing an exam and failing it miserably.

HB: So, sleep preps the brain for learning appropriately.

MW: Yes. And you can look at both sides of that coin. You could ask, *"What is the detriment to your capacity for new learning when you haven't had sleep?"* but you can also ask, *"What is it about sleep, when you **do** get it, that seems to restore and refresh that learning capacity?"* And we've been able to answer both of those questions.

Firstly, we know that when you take sleep away, when someone has not had sufficient sleep the night before, there's a very specific structure in the brain called the hippocampus, which is the quintessential reservoir for where you create new fact-based, textbook-like memories—

HB: Where is the hippocampus exactly?

MW: You have two of them, just as you have two of almost all brain structure: one on the left and one on the right. It's almost like a long, slightly bent cigar that starts at the back of the head and curls down around to the front.

We know, for example, that if you damage that structure, or if that structure is surgically removed due to conditions like epilepsy, you become densely amnesic: you can no longer form any new memories.

And it's that very same structure that sleep deprivation attacks, blocking your brain's capacity for new learning. By the way, we don't know if that's reversible. We don't know how quickly, if at all, your hippocampus can recover after severe sleep deprivation. That's one of those interesting questions.

But let me try to add some brightness to the story. When you're getting good sleep, how does that work? What's the mechanism of benefit? This brings me back to those sleep spindles that I was talking about before. What we've found is that the amount of these sleep spindles that you're having at night seem to accurately predict the degree to which your hippocampus is refreshed and restored in terms of its renewed learning capacity the next day.

HB: So you are able to measure these sleep spindles over time?

MW: You can measure them over time and you can measure them over the brain as well. It seems to be the case that the more of these sleep spindles you have—especially over the frontal part of the brain, which has direct connections to the hippocampus—the greater the degree of refreshment of your learning the next day.

HB: I'm not a biologist, and I often find myself vainly searching for some bigger-picture understanding when people show me correlations with different phenomena in different regions or whatever, because I don't have any clear sense of what, exactly, is going on. So let me back up. I understand what it means to refresh something of course. And presumably there's a lot of activity going on in my brain in this particular region and you can tie that empirically to people's ability to remember things, to learn things and so forth—hence you have some empirical justification for this notion of "refreshment". But then I ask myself, *"What's actually happening here? What's the actual mechanism?"* You must have some theory, some idea of what's going on at the underlying levels.

MW: We do. There's good evidence now. If you combine the studies in humans with studies in rodents, for example, you get an interesting picture. I'll start with an analogy and then I'll overlay biology on top of that.

The analogy would be that the hippocampus is like a USB stick: it's very good at grabbing information on the fly from different sources. But, just like a USB stick, it has a limited storage capacity, whereas your cortex—this mass of tissue on top—is essentially like the hard drive with a vastly greater storage capacity.

What we think happens is that during sleep, after you've acquired lots of information during the day on your hippocampus USB stick, the hippocampus will then essentially have a therapy session with the cortex and they'll have a dialogue during your sleep, during which time the hippocampus, your USB stick, will upload its information to the cortex, the massive hard drive storage space of your brain.

There are two benefits to this dialogue, this therapy session, which will nicely bring me to the second of the three functions of sleep in terms of memory, which is protecting and saving the memories that you learned the day before.

Firstly, information that was on the hippocampus, the USB stick, gets put on the hard drive and now it's safe in a larger storage capacity.

And secondly, when you wake up you've cleared your hippocampus USB stick and now it's ready for more information. So what you learned yesterday is safely on the hard drive, and your hippocampus—your USB stick—is now ready for new learning the next day. There's this beautiful reciprocity that happens.

How is that happening? What we know from animal studies, and now from some evidence in humans, is that in the hippocampus, when you're learning specific information, there are signature firing patterns coding what you're learning. There's a signature of memory nested in the firing cells mapping that information, and you can measure that.

Let's just say that you can add a sound to each of those cells and maybe—as a rat is learning a maze, say, you're recording from those cells of the hippocampus and you'll hear the sound of the memory trace being laid down.

What's remarkable is that when they go into NREM sleep—when you're getting these sleep spindles and these slow waves—if you were to turn the recording device back on, you would start to hear the sound of the same signature firing pattern. What's strange is that it's sped up by a factor of perhaps almost 20. What we believe this is reflecting is exactly the replay of the memory traces from the hippocampus up to the cortex. So there is really quite good biological evidence for this type of a therapy-session dialogue actually happening. I'm using an analogy, of course, but there's actually hard, neurobiology underlying it.

I've already essentially started to describe the second benefit of sleep, which is that you not only need to sleep before learning to get your brain ready to acquire that information by clearing out the USB stick, but once you've filled the USB stick up, you also need to sleep after learning to then grab a hold of that new information and solidify it and cement it into the neural architecture of the brain.

As I said, what we believe is happening is a transfer process: taking information from the fragile, vulnerable state of the hippo-campus USB stick up into the more permanent, solid situation which is your cortex, your hard drive. That also seems to depend on deep, NREM sleep. As a consequence, if you sleep after learning, those memories are not going to be deteriorated by the ravages of time. You're not going to forget over time.

However, if you remain awake throughout the night, if you pull another all-nighter, say, the memories that you have—because you don't have the chance to go to sleep and have this dialogue and trans-fer—they will degrade, so you'll get "accelerated forgetting".

That's the second function of sleep. Sleep will take things that you've just learned—and if you've had a good night of sleep before, you'll have had efficient learning to begin with—and after learning, sleep will grab those individual memories and strengthen them.

That was probably the view of sleep and memory until about five or six years ago, until we started to realize that there was a *third* bene-fit. This third benefit—which may actually relate to REM sleep—is that sleep also seems to be far more intelligent than simply grabbing individual memories and saving them; sleep actually seems to be able to take different types of information and start to *interconnect* them and *coordinate* their association, so that you create patterns of more general knowledge.

For example, we know that memories do not sit like an isolated island within your brain; they would be profoundly useless like that. Memories are instead richly interconnected in these webs of association. How and when does the brain decide to build those associations? Which associations should it create and which should it not? When we're awake during the day, that's a time when the brain starts to understand that one thing should be linked to another. Sleep also seems to be a time that the brain does this, but in a slightly different way.

I think what's happening—we've got some emerging evidence now—is that during the day when you're awake, it's the obvious, direct connections that the brain starts to identify and create. During sleep, it's as though you go to the other end of the spectrum. It's as though you start to fuse things together that, at first, really *shouldn't* normally go together. But it turns out that occasionally, when they *do* go together, they cause marked advances in evolutionary fitness, the basis of memory-biological creativity.

It's almost as though, when you're awake, it's a Google search gone right: you input "coffee mug" and the first page is a link to Amazon and other places where you can buy coffee mugs. But during sleep, and perhaps during REM sleep and dreaming, when you start to try these associative algorithms, now the first hits that you get are page 20, which is about some field hockey game in Utah. Where did *that* come from?

It turns out that if you look into it, there actually *is* a very strange, bizarre pathway connecting these things. What seems to be happening during sleep, and perhaps during REM sleep, is that you're testing out associations. It's a little bit like memory pinball: you're taking what you've learned during the day and you're launching it up into this attic of everything that you've learned before in your past experience, and you bounce it around these nodes.

HB: This seems to make sense to me. Maybe I'm just easy to convince, but it seems to make sense because during the day you don't have as much processing power to devote to this sort of thing. You have all sorts of other things going on. You have to stay focused. You have to, to take your analogy, be looking at page one of Google.

But when you're relieved of all of those things, when your brain doesn't have to be focused with all of that, you have the luxury of being able to go all over the map, which presumably involves a little more time, making more connections from an evolutionary perspective, maybe taking long shots.

Even just anecdotally what you're saying seems to be reinforced by a very common experience that most people have had: this whole idea of "sleeping on something," together with the sense that you can come up with some really interesting ideas while you're sleeping.

I think most people have had this experience: they wake up and they have some good ideas. This seems to be very much in keeping with that possibility.

MW: Exactly. When was the last time anyone ever told you, *"You should stay awake on a problem"*? There's a reason for that.

I like your thought; and it reminds me that when you're awake, it's about reception, while when you're asleep, it's about reflection. Reflection is not just simply taking stuff that you've learned and hitting the save button; it's also about starting to try and understand what you've learned.

HB: It's a creative process.

MW: Exactly: it's a creative process. It's not just simply knowledge, which is holding on to what you've learned. It's about wisdom, which is knowing what it all means when you fit the pieces together. We're now starting to understand that this is a third, and perhaps ultimate, function of sleep.

You're right about the link between sleep and creativity. Just within science there are innumerable examples of classic scientific discoveries that have come by way of dream-inspired insight.

Dmitri Mendeleev put together the periodic table of elements, the fundamental elements of our universe, by way of sleep-inspired insights.

Otto Loewi won a Nobel Prize for discovering neural transmission, how chemicals are released from one brain cell to the next. And he designed those experiments in his sleep, if you believe the story.

And it goes beyond science, of course. Paul McCartney wrote all sorts of wonderful Beatles tunes: *"Yesterday," "Let It Be,"*—

HB: Right—this is your whole Beatles obsession. I knew you'd bring it up eventually. I'd been warned.

MW: In every interview I do I have to try and give a shout-out to one of my brethren from Liverpool.

HB: One minute you're talking about a Nobel Prize, and then all of a sudden—wham!—Paul McCartney.

MW: It was seamless, wasn't it?

But to get back to what we were saying, it's clearly true that there have been many past examples of how people use what we've been describing: the creative power of sleep.

One great example is Thomas Edison—who, as it turns out, has a lot to answer for, in terms of how we're sleeping. People have often said to me, *"Well, Edison, this brilliant, creative guy, he was known to be a short sleeper."* Now, there's some argument as to whether or not that's true, but even if it was the case that he was a short sleeper at night, he was a habitual napper during the day. In fact, he would use naps for creative insight.

It was genius. What he would do—and this is such a mark of the man—is that he would have a piece of paper and a pencil next to him, and place an upside-down metal saucepan underneath the armrest of his chair. Then he would take two steel ball bearings in his hand, rest the back of his arm on the armrest of the chair, and he would slowly start to fall asleep. At a certain point his muscle tone would relax sufficiently so that he would release the steel ball bearings and they would crash down onto the saucepan, which would wake him up, and then he would start to write down all of the ideas that he was getting from sleep. Absolutely brilliant.

Clearly some people have actively used sleep as a tool for creativity. In terms of sleep and information processing, now we understand that sleep is vitally necessary for many steps of memory processing:

the formation of memories, the holding on to memories, and how to interconnect and stitch those memories together so that you can understand the big-picture rules and the gist of what we call this world that we live in.

Questions for Discussion:

1. In what ways could an improved understanding of the biology of sleep help improve our formal education systems and practices?

2. Does this chapter make you reconsider the efficacy of "pulling an all-nighter" to study for a test?

3. How might future sleep research shed valuable light on the nature of the creative process itself?

V. Sleeping Better?

Pharmacological effects and self-improvement

HB: I'm sure we're at far too preliminary a stage to say anything comprehensive about this, but you must have some suspicions as to how best harness this knowledge to our advantage. I understand everybody is a little bit different, but if I'm aware of these basic processes that you're telling me about, can I make any sort of judgement about how would that affect me in terms of how I might go about approaching my own sleep cycle, whether I'm sleeping too much or whether I'm not sleeping enough?

You mentioned napping. Are you at a stage where you can say, *"You might want to try sleeping for this amount of time each day,"* or, *"You should go see your sleep therapist to find out what your ultimate profile would be"*? If I want to be as productive as I possibly can, if I want my memory to be as sharp as possible, if I want to be as creative as possible, how can I go about doing that in terms of working on my sleep?

MW: What we currently know is that once you start to get less than seven hours of sleep a night, you can start to measure impairments in people's function. The reason that I bring that up is it suggests that somewhere between seven-and-a-half to eight-and-a-half hours of sleep a night for every average, adult human being is about the optimal amount.

One of the frightening things that seems to happen with insufficient sleep is that your subjective opinion of how you're doing with too little sleep is a miserable predictor of how you're doing objectively.

This is that classic case of people saying, *"I can survive on six hours of sleep. Sure Doc, I know you tell me all this stuff about sleep,* but *I'm* one of those who's okay with just six hours," to which I say, "*No, I know that you* **think** *that you're okay, but you're actually not.*"

The analogy would be the guy at the bar who, after six shots of vodka, picks up his car keys and says, "*I'm fine. I'm completely unimpaired. I'm totally alright to drive.*" He may *think* he's not impaired, but it's a dangerous situation, and the same danger underlies insufficient sleep.

HB: That reminds me of something very similar that came up when I was talking to your colleague Stephen Hinshaw (see the Ideas Roadshow conversation *Understanding ADHD*) when he was describing the effect that non-ADHD patients had when they took ADHD medication like Adderall, so-called "neuroenhancers". They were convinced that the medication had this wonderful effect and that taking it resulted in them doing much better on the test, but when you actually looked at the test results, it turned out not to be true.

So people *think* they can survive on six hours of sleep, but most people actually can't. But what about napping in between or somehow staggering that time? Is it so important to have seven-and-a-half full hours of sleep? Could I break it up into two-and-a-half hour segments? Churchill did that, didn't he? Wasn't he supposed to have slept every six hours for two hours, or something like that?

MW: Churchill certainly had a very erratic sleep profile. He also suffered from significant, major depression and that's probably one of the reasons for his underlying sleep issues.

HB: So perhaps he's not the best example, then.

MW: Right. But there *is* currently a debate about how, exactly, we should be sleeping, naturalistically speaking, about how we were designed to sleep, or how we evolved to sleep.

In industrialized nations we currently sleep in what I would call a monophasic pattern: we have one bout of long sleep at night and then we're awake for sixteen hours, so that we should be sleeping for eight hours in each twenty-four-hour period.

There is an argument that we should perhaps be sleeping biphasically: a somewhat long bout during the night, maybe six-and-a-half or seven hours, and then an afternoon nap, that sort of siesta-like Mediterranean behaviour. Why would I argue that? Well, it turns out that we can measure people's biological alertness and also drops or dips in their alertness that will send them towards sleep.

There are a variety of specific measures that you can use to assess this, but for the moment let's just say that they exist. What seems to happen is that as you're going throughout the day, your alertness, attention, and your physiological desire to stay awake start to deteriorate. It will really start to drop down, for most people, around about ten, eleven, midnight. Then, as you get through to the early morning hours, that activation within your nervous system starts to rise up again and you start to wake up. By eleven or twelve o'clock midday, you're now nicely alert.

What's strange, however, is that once you get to two or three in the afternoon, you start to come down again and then you'll come back up around five or six.

HB: That's just so true.

MW: Everyone's had the experience of being in a warm room after a big lunch, they're at the big meeting table, somebody's giving a presentation...

HB: *Especially* when someone's giving a presentation.

MW: Especially if it's *me* giving a presentation.

And you'll see these interesting sorts of head nods. It's not because these people are listening to good music: they're falling prey to this dip in their alertness, which would argue that we're biologically predisposed to go into this sleep pattern in the middle of the afternoon.

You have to be careful with naps though, because they're a double-edged sword. One of the ways that we develop tiredness or sleepiness is that the longer you're awake, the more that a certain chemical builds up in your brain, which forces it to go to sleep. That chemical is called adenosine.

It's a by-product of cellular metabolism. The more of it you have, the sleepier you feel; at a certain point it gets so potent, in such high concentrations, that you can't resist sleep and you fall asleep.

Then when you're sleeping that adenosine is removed from your system. Think of it like a pressure cooker: it starts to build up and this pressure to sleep develops and then when you go into sleep, the pressure valve is released.

HB: How does the adenosine get removed? Physically, what's actually going on?

MW: What we find is that this deep, slow-wave sleep I was talking about earlier seems to offer a biological mechanism to clear out the adenosine at the cellular level. By removing it and degrading it, you start to take away that sleep pressure and the brain naturally rises back up to wakefulness, which is going to happen late in the morning.

The reason that naps can potentially be a little bit dangerous is that, let's say, you wake up at seven or eight in the morning and you start to build up these concentrations of adenosine across the day. You're on a nice trajectory to fall asleep at ten or eleven at night, but then you take a nap in the middle of the day. Naturally, you'll dissipate some of that sleep pressure so it goes down again. Then, as a consequence, come bedtime at ten or eleven, you're not going to feel as tired anymore. As a result, you then stay up until maybe one or two in the morning which means, come seven or eight, you'll want to sleep for a little longer, but unfortunately your workday forces you to get up too early.

So you have to be a bit careful. Taking naps can cause issues with insomnia—not with all people, but some. Naps are interesting. It's almost an anthropological question of how we should be sleeping as a society.

The idea that we should force ourselves awake in the morning rather than letting our brain do it naturally is utterly artificial. If you speak to cultures that haven't been touched by electricity, do they have alarm clocks? No. It's profoundly odd. The first alarm clock was the factory whistle; that's how society began waking up en masse. It's an odd thing that we artificially wrench our brain out of something that it desperately needs. No other species does that.

HB: This brings up another topic. When you talk about artificiality, that makes me think about people who take sleep medication and tranquilizers. Would that have the same effect as these natural processes that clear out adenosine?

I could imagine that it might not, because I've heard people say things like, "*Well, I was asleep, but it wasn't the same sort of sleep. It was a chemically-induced sleep.*" So I'm imagining that even though they are, to all intents and purposes, sleeping just like everybody else would be sleeping, different processes are actually going on in their brain and they don't have this cleansing process to the same extent that somebody who fell asleep naturally might.

Is that true, or is that not the case?

MW: That is perhaps one of the most fundamental questions that has not been thoroughly answered yet, "*What is the effect of sleep medications?*" The reason it's so fundamental is that, if you look at the numbers, millions of people every single night, just here in the United States alone, are utilizing these sleep medications.

Do we know that they change the architecture of sleep? Yes, they do. These common sleep medications—I won't name them, but many of them that are currently being used today—don't seem to give you the depth of deep sleep that you would normally get. They may also reduce the amount of REM sleep that you get.

Back 15 years ago it was even worse. The old drugs that we used to call the sedative-hypnotics—drugs like benzodiazepines, tranquilizer-like drugs—would definitely stop you from being awake. Nobody would argue that you were awake when you took those drugs. But equally, to argue that you were asleep would actually be very difficult based on the patterns of brainwave activity. You were simply sedated.

It's the same case with alcohol. People think alcohol is a useful way to get to sleep if they're having sleep problems. Alcohol doesn't induce sleep—alcohol sedates you. And even worse, alcohol will actually *reduce* the amount of REM sleep that you have and fragment your sleep.

HB: When you're sedated, what does that mean in terms of brain-waves? You have the same activity just modulated somehow? What does "sedated" actually mean?

MW: Essentially it means that the brain is not showing patterns of normal waking electrical activity, but it also isn't necessarily a pattern that is prototypical of sleep either. It's a different pattern of electrical activity. Those deep, slow-wave patterns that we spoke about before —nice, deep, naturalistic sleep—they start to become altered by these types of sleep medications. How much of that deep sleep you have, and particularly the electrical quality of that sleep, starts to change as a function of taking these drugs.

I think one of the important questions, especially now that the prescription age of those drugs is starting to decrease, is, "*How is that impacting the normal functions that we now know sleep seems to serve—like learning, memory, and emotional regulation?*" We currently don't have good answers to those questions.

There may be wrestling matches over how those studies are funded. The drug companies are not eager to fund those studies because what would happen if we found that people are sleeping for eight or eight-and-a-half hours but they're not getting the benefit? Even worse, what if they're actually getting deterioration?

I know of one study by Dr Marcos Frank and his colleagues that was looking at how the brain seems to rewire itself during sleep. We know now that the brain can also make new connections. It's not like a hard-coded circuit board.

The brain is very plastic and that plastic reorganization, based on what we've learned during the day, happens during sleep. But what they found was that if you dose rats or other animals with these types of drugs, even though they seem to have sleep, that re-wiring doesn't take place—there isn't that nice, normal plasticity that helps the brain with learning and memory. If anything, in fact, it was actually *impaired*—it wasn't simply that it wasn't happening.

That's one study that I know of. It's a remarkable study and I think it's a frightening demonstration that perhaps the electrical and the physiological quality of our sleep is so non-naturalistic on these medications that they not only prevent us from getting the benefit, they may actually be detrimental.

Now, there's much work to be done. We don't know if that's necessarily the case. I'm not trying to scaremonger or drive people away from these medications.

HB: But that's all the more reason to conduct the kind of research that you're doing.

MW: Yes, my intention is simply to motivate these sorts of studies.

Also, I think people should be aware that you don't necessarily need pharmacology to improve your sleep. Even for severe insomnia, there are now very good behavioural techniques. One is called cognitive behavioural therapy. It's the first-line choice, a non-pharmacological choice. And it's remarkably efficacious.

HB: What is it? How does that work?

MW: People take several days to learn these techniques. It's a collection of individual factors that help alter your behaviour so you can get good sleep.

For example, not being in the bedroom while you're awake, not lying there for too long and being awake if you're not falling asleep, because then the brain starts to associate being in bed with wakefulness, not sleep. So get up and get out of the bedroom.

You can also try to short-sleep people, which might sound counter to what you're trying to do. But if you reduce the amount of sleep that they're getting, their adenosine starts building up. Then on subsequent nights, they are able to fall asleep far more easily, and they start gaining confidence and psychological reinforcement that they can sleep.

HB: So they learn how to sleep, effectively.

MW: Exactly. These and many other features are built into this program called cognitive behavioural therapy. It seems to create not only naturalistic sleep, but also long-lasting effects that these drugs, once you stop taking them, don't provide you with.

Again, that's not to suggest that pharmacology can't be useful for short-term insomnia. But for chronic insomnia, cognitive behavioural therapy, a non-pharmacological treatment, seems to be very efficacious.

Questions for Discussion:

1. *Are you concerned about Matt's comments that* **"The drug companies are not eager to fund those studies because what would happen if we found that people are sleeping for eight or eight-and-a-half hours but they're not getting the benefit?"** *What does this imply about the role of corporate funding in science? To what extent do you think that our current situation parallels smoking and cigarette manufacturers in the 1960s and 1970s? Under what circumstances does someone like Matt have to be very careful about how he expresses himself publicly on these issues?*

2. *To what extent is the line between "natural" and "artificial" treatments fuzzy given that the underlying mechanisms of brain biology are necessarily chemical in nature?*

VI. Emotional Regulation

How sleep helps keep us balanced

HB: Earlier you spoke about how, in terms of brain function, the impact of sleep seemed to fall into two large areas, learning and memory and emotional regulation. You've spoken at length about the different ways that sleep impacts learning and memory, so now I'd like to move on to the second category of emotional regulation. I don't know what that is, actually, so before you tell me how sleep affects it, perhaps we should start there.

MW: Emotional regulation simply means that we're able to deploy our emotions effectively and control our emotions effectively. If you think about people who are irrational or excessively emotional, they seem to make non-optimal choices because their emotions just sort of overpower their rational, logical, decision-making.

But the inverse can be true too: people can suffer from a lack of emotion. Think about depression. One of the hallmarks of depression is something called anhedonia, which is a fancy term that simply means you can't gain pleasure from normally pleasurable things. That's one of the defining features of depression. So that is a case where there is not enough emotion happening.

Emotion, unlike learning and memory—which is a linear function; the more of it you have, the better presumably—emotion is sort of an inverted-U function. Too little is not good, just enough is optimal, and too much is also not good.

What we're now understanding is that it seems that as long as you're getting enough sleep, particularly REM sleep, you'll be placed at the sweet spot right at the top of that inverted-U so you can control your emotions optimally: you don't have too much, but you also have sufficient amounts.

HB: What evidence do we have for that, exactly? How does that work?

MW: We've explored this question by again using this bi-directional approach: you can dial sleep down—taking it away with deprivation, and seeing if you can trigger an amplified emotional reaction from the brain as a consequence, or you can give sleep back and sort of dial it up—maybe even insert sleep where it normally wouldn't be by using a nap during the day—to see if you can then cause this nice, palliative amelioration of emotion reactivity.

We look closely at a particular centre within the brain, a region called the amygdala—you have one on the left and one on the right. The amygdala is sort of the kingpin of your emotional reactivity. And it turns out that when you are sleep-deprived the amygdala becomes excessively reactive: that emotional centre of the brain is amplified in its reactivity towards negative experiences.

So, for example, we take people and either sleep-deprive them or give them a good night's sleep. Then we place them inside the fMRI scanner and we'll show them increasingly negative and unpleasant images to see how their brains will react.

When you've had a good night's sleep, it's not as though your brain isn't reacting to these images—it is, and it should be—but it's reacting in a controlled way. There's a moderate reaction. But in those people who are sleep-deprived, you see an excessive reaction. The emotional brain is more than 60% amplified in terms of its reactivity to negative experiences, without a night of sleep.

HB: This is the signal itself you're looking at or—

MW: Exactly, it's the signal within this structure, the amygdala. You can look at it with these MRI images and you can assess how strongly, how brightly, it's burning in reaction to these images. You can quantify that empirically, and that's how you can measure the difference and obtain a number like 60% in terms of an amplification.

What's also interesting is the question, *"Why is this deep, emotional brain centre so excessively reactive without sleep?"* Using some additional analysis you can demonstrate that, when you *do* get a good night's sleep, there's another part of your brain, in the middle part of the frontal lobe that's strongly connected to the amygdala, this deep, emotional centre.

Why is that important? It's important because this part of your frontal lobe—the frontal lobe is like the CEO of the brain: it makes high-level, executive decisions—seems to be regulating the amygdala with negative inhibition. In other words, the frontal lobe is the brake to your emotional gas pedal.

HB: Right. So it tells your amygdala, *"Don't worry about this. It's okay."*

MW: Yes. Or rather, *"Worry about it, but only to a certain degree."* So there is a nice, balanced mix between the gas pedal and the brake. But without sleep, that frontal lobe region, in terms of its connection, is severed from the amygdala.

So, without sleep you become all emotional gas pedal and no brake. That same type of profile of abnormal brain activity has also been associated in conditions like post-traumatic stress disorder, or PTSD.

So that's the bad stuff that happens when you take sleep away.

Then you can turn things around and ask the question, *"How exactly is that connection between the gas pedal and the brake improved when you **do** get sleep?"*

And what we've found is that, unlike learning and memory which seem to be related more to NREM sleep, emotional processing and mental health do seem to be directly related to REM sleep.

One of the benefits seems to be that REM sleep can take emotional experiences that you've had the day before and essentially act like a soothing balm. During REM sleep these emotional centres of the brain become active once more, and we believe that allows you to bring those emotional experiences from the prior days back into play, as it were.

There's a remarkable feature about REM sleep. There is a particular stress neurochemical within the brain called noradrenalin or norepinephrine. If we were to experience an earthquake right now, for example, this chemical would spike in our brains and it would help burn emotionally difficult or traumatic memories into the circuits of our brains.

During REM sleep, despite the fact that those same emotional centres in the brain are reactivated, that stress neurochemical is suppressed. What we believe this offers is a chemical cocktail that is perfect for, as I was saying earlier, a sort of "therapy session". It's as though you can reprocess these traumatic memories, but in a brain state that's devoid of any stress chemistry.

So it's in this way that we think that REM sleep acts like a soothing balm: you can bring these emotional memories back to mind—and perhaps this is partly what dreaming is about too—but you can start to strip away that bitter, emotional rind from the informational orange of the experience, so you can essentially "deal with it."

Of course, the reason that you have emotions when you're awake and learning is so the brain can red-flag and prioritize salient experiences, so it's important to have emotions at the time of learning. But what we've argued is that it's not good to hold on to that stress-blanket of emotion that's around the memory long-term, because that way what you get is a state of chronic anxiety in your autobiographical memory networks.

We think REM sleep offers a perfect environment to strip away that emotional blanket from the core of the experience, so that after several cycles of sleep, or several nights of sleep, what you come back with when you wake up is a memory of an emotional event, but it's no longer emotional itself.

HB: It's like the brain is, in a sense, tagging that event so that it can reprocess and analyze it later during REM sleep.

MW: Exactly. It's as though it raises the red flag of importance during wakefulness and tags the memory: it's saying that the information of that experience is important by packaging it in this wrap of emotion.

But that emotion itself isn't beneficial, from an evolutionary perspective, to maintain long-term. If that was true, every time you recollected important things from your past, not only would you be able to recollect the information, you would also regurgitate the same visceral emotional reaction that you had at the time of learning, which is not good.

HB: That's not good. And it's terribly wasteful as well.

MW. Right—it burns energy, which is always a dangerous idea, evolutionarily-speaking.

The quintessential disorder where this mechanism fails is post traumatic stress disorder, PTSD. What you will hear when you speak to these patients in the clinic is that they simply can't "get over" the event. When they're cued or triggered to recollect that traumatic memory—imagine a war veteran hearing a car backfire in the grocery store parking lot, say—not only do they have a flashback of the memory of the trauma, but they are also forced to relive the whole thing *emotionally.*

HB: They experience the same chemicals, the same emotion, the whole thing.

MW: Exactly. And what that tells me is that the brain has not been able to strip away and divorce the toxic emotion from the memory. I don't think it's coincidental that one of the diagnostic hallmarks of PTSD is repetitive nightmares and bad sleep. It's as though the brain is offering up this highly-charged, emotional memory and saying, *"Please REM sleep, do your elegant trick of carving off the emotion from the memory,"* and this process fails for reasons that we're only now just starting to understand.

So the next night, what happens? The brain comes back and says, *"I've still got this highly-charged, emotional memory. Please strip away the emotion from the memory,"* and it fails again, like a broken record, which is quintessential of the repetitive nightmares that are present in PTSD.

Questions for Discussion:

1. To what extent do you think that the common advice to "sleep on it" is a tacit recognition of the role that sleep plays in emotional regulation, helping us avoid making impulsive, emotionally-driven decisions?

2. How much of the notion of "emotional regulation" is solely focused on negative rather than positive emotions? What role might positive emotions play in this process? (Those interested in this topic are referred to the Ideas Roadshow conversation **The Science of Emotions** *with UNC social psychologist Barbara Fredrickson.)*

VII. Sleep and Aging

Grappling with the inevitable

HB: There was something I read that deeply depressed me.

MW: My research often does that to people.

HB: Well, I'm referring to the fact that, as a middle-aged guy, it seems that my prefrontal cortex is shrinking at some alarming rate. I'm also more sensitive to jet lag and I'm not nearly as good at dealing with all sorts of other sleep-related issues. I think most people appreciate the fact that, as they get older, their sleep patterns change. You don't have to be a sleep researcher to be aware of that.

But what's going on? Why is this happening to me as I'm getting older and is there any way of combating this?

MW: Trust me, I'm right there with you. I may be just a few years younger, but it's in the mail for me. As I described earlier, we know that deep, NREM, slow-wave sleep—which helps with the saving and hence remembering of facts that you learned the day before—that type of sleep shows remarkable decreases across the lifespan which starts to happen quite early on.

But we don't yet know the answer to two additional questions: First, *Why does your sleep, particularly that deep sleep, start to degrade so rapidly as we get older?* And secondly, *What is the functional consequence of the degradation of that particular type of sleep?*

We've known for a long time that, as we get older, our learning and memory get worse—that's the quintessential cognitive feature of getting older. But what we also know from physiological sleep studies is that a biological hallmark of aging is worse sleep. Now that we know sleep and memory are related, it raises the question, *"Are those two features of aging—bad sleep and bad memory— simply independent, or are they significantly interrelated?"*

We recently published evidence which argues that these aren't simply co-occurring phenomena; they may actually be significantly interrelated. What we found was that there is a part of the brain that seems to correlate with your ability to generate this deep, slow-wave sleep. That is the frontal lobe, just as you've described, in particular inner parts of the frontal lobe. As we get older, our entire body, including our brain, deteriorates, but some parts of the brain will deteriorate more rapidly than others during the aging process.

One of the parts that degrades with aging at an accelerated manner is the frontal lobe, and these inner parts of the frontal lobe in particular. It's exactly that part of the brain that seems to be critical in helping us generate deep sleep. And what we found is that, as one gets older, the degree to which this part of their brain has deteriorated—what we call atrophied—directly predicts the amount of reduction, or the loss, of deep, slow-wave sleep. As a consequence, we then predicted the degree to which this person will forget, rather than remember, facts that they learned during the previous day, after a night of sleep.

In some ways, it's a very depressing story—

HB: In **some** ways?

MW: Well, I'll try to give you a silver lining in just a second.

This brain deterioration negatively affects your sleep, and consequently the functions that rely on those different types of sleep are also cast off as your sleep deteriorates.

I should back up. I'm not suggesting that aging is a sleep disorder. It's clearly not.

HB: Of course not. But these things are correlated, as you said.

MW. Right. It's a multifactorial problem. However, it's becoming apparent that bad sleep is an often under-appreciated factor that is contributing to what we call cognitive decline.

The reason this is exciting news is because many of the other factors that we know contribute to aging—changes in blood flow, changes in the white matter tracts that connect different parts of your brain together, which do change with aging—these are all fiendishly difficult to treat. We actually don't have good treatments to change blood flow or restore structures within the brain. But sleep, if it's a contributing factor, is a modifiable factor. We **do** have ways to improve sleep.

So now we have a target that we can try to improve; and, as a consequence, restore back some degree of memory function, the degree of memory function that sleep is supporting. We're now moving towards studies that do precisely that: try to restore sleep and consequently memory function.

HB: Are there existing studies, or will you be starting studies, with older people to embark upon sleep therapy programs to see if cognition or memory is affected in any way? Is that the next step?

MW: It is. Based on that work about aging that I just described, there are two next steps: One is to look further into the aging process and dementia, particular in terms of Alzheimer's disease where sleep disruption is rife and remarkably apparent.

HB: So this is closing the 14-year loop that you were talking about earlier.

MW: Exactly. I'm finally coming back to it after 14 years.

The other avenue is to try and somehow restore sleep in elderly people. We're applying for grants to do both of these streams of work. How do you restore sleep? Well, you can think about it pharmacologically or you can think about non-pharmacological ways.

Another interesting way to restore sleep, one that we're going to try and approach, is actually by using electrical brain stimulation.

I should stress that this is not like *One Flew Over the Cuckoo's Nest*. It involves applying electrodes to the front of the brain and seeing if, by using tiny amounts of voltage—you don't even feel it—you can boost the size of those big, slow brainwaves that start in the frontal lobe. These electrodes essentially stimulate the natural rhythm of the brain. The idea is that this stimulation will increase the size of these brainwaves and this should consequently increase the quality of that sleep. If you can causally increase the quality of that sleep, you will like be able to causally enhance memory.

This has already been demonstrated in young, healthy people. Even in their sleep, you can actually electrically, artificially amplify their brainwaves and you can almost double the amount of benefit.

HB: Why are you wasting your time with young, healthy people? Help the older guys.

MW: You might think that every night I apply these electrodes to my own head. But I would argue that people should get naturalistic normal sleep and let evolution do its job. It's taken hundreds of thousands of years to develop this complex system.

But this is a therapeutic, intervention possibility. We can use this and other mechanisms to try to restore healthy sleep in older people, or at least make it as healthy as possible. And by doing this we may be able to restore some of that cognitive function that is lost as a consequence of bad sleep.

Questions for Discussion:

1. To what extent might regular exercise serve the added, substantial benefit of helping us sleep better as we age?

2. Why does Matt interrupt himself and say, "I'm not suggesting that aging is a sleep disorder"? How is this related to the phrase "Correlation is not causation"?

3. Would you be willing to try electric brain stimulation to improve your sleep if you had the chance?

VIII. Sleep Stigma

Lazy thinking

HB: It seems to me that most of the time, sleep is not just under-appreciated, it is not appreciated at all. In the past, I must confess, I've been a prime example of this. I've often asked myself, "*Why am I wasting a third of my life sleeping?*"

MW: That's the funny thing about sleep: it's got this awful stigma in society. Nowadays it's almost as though people associate sufficient sleep—and I choose that word very carefully—with laziness. We as a society have this skewed view of what sleep is. We didn't always have that view.

Nobody looks at an infant sleeping during the day and says, "*What a lazy baby!*" Why? Because we all realize that sleep for that baby is absolutely essential.

But quite soon after that, we abandon the notion of sleep being useful and, if anything, we are actually against it. We wear our badge of sleep-deprivation like some sort of honourable emblem and we're proud to say how little sleep we get.

HB: But there *is* a bit of laziness involved too, right? We've all had the experience of waking up and thinking, *I could get up, but I'm just going to go back to sleep for another hour or two*. I'm not talking about getting six hours of sleep when you should have had seven. I'm talking about getting the recommended amount of sleep and basically feeling like you could get up, but deciding to sleep longer.

MW: What you're describing is something called a sleep debt. If you *could* go back to sleep, that tells me that you've not been getting sufficient sleep beforehand.

Some people will need nine hours of sleep while others will need seven or seven-and-a-half. But there is this mistaken notion that when your brain needs sleep, you're being lazy. It's a biological necessity.

Human beings are one of the few species that deliberately deprive themselves of sleep. The rest of the animal kingdom doesn't do that. If they need to sleep, they sleep—and they will be better off for it.

It's a very strange situation and we need to do away with that stigma. Sleep is the third pillar of health along with exercise and diet. I think we now have enough science and people are starting to convey that message. Habits may finally start to change.

Questions for Discussion:

1. Have you underappreciated the value of sleep in the past? Has reading this conversation changed your attitudes towards the importance of sleep?

2. How can we best ensure that people are made aware of the importance of sleep to living a healthy life?

IX. Further Questions

Motivation, narcolepsy and vicious circles

HB: I have a specific question about one of your studies that I was reading about where the effects of sleep on people's memory was being tested. There was a list of words—in this case I believe it was both real words and nonsense, made-up words—and the idea was to test the ability of subjects to recall these words on the list before they went to sleep and compare it with their level of recall after they slept, or something like that.

But when I try to remember something, a key factor for success is often whether or not I want to remember. For me there is a significant act of volition involved. That seems to be left out of these kinds of experiments. I can well imagine thinking to myself if I were a subject, *Well, who cares about these crazy words?*

MW: What you're describing is that motivation normally precedes learning. The reason that we try to form new memories is often because there is motivation involved—not always, but often. There are ways to model that, and we've done that with some of these sleep studies. It turns out you seem to get an even greater benefit when you motivate the subjects. How do you do that?

What you can say is something like, *"For every one of these word pairs that you are about to learn, I will give you 50 cents if you remember it."* Of course you can modify the amount until you reach a point where a significant amount of motivation occurs. You can give $1, or $2, say.

HB: This is America, after all.

MW: Exactly: classic capitalist mentality. So let's incorporate that into designing our studies. You can essentially cue each learning trial with different degrees of monetary reward.

Now it becomes a little bit more like what you're describing, which is that the subject is now motivated to learn. So you can do studies like that to take that into account.

HB: OK, that answers that. My next question is about narcolepsy. What's going on there? What's happening to these people who are suddenly falling asleep without any provocation?

MW: Narcolepsy is a serious condition. It's about as common as multiple sclerosis: about 300,000 people suffer from narcolepsy. It is a disorder that, at least at one level, is due to a lack of a specific chemical in the brain called orexin.

You might imagine the brain mechanism that can make you awake or push you back into sleep as sort of like a seesaw. One of the chemicals that helps force your brain on into wakefulness—and when it's removed puts you back into the default of sleep—is this chemical called orexin.

You can think about narcolepsy like a faulty light switch. If you speak to an engineer they'll tell you that the way any switch should work is that when it's on, it's fully on, and when it's off, it's all the way off: it's a binary gate.

And orexin, when it's in high concentrations, allows you to knock that switch all the way on or all the way off. Narcoleptic patients, however, have a biological deficiency in this chemical—and as a consequence, their sleep-wake switch hovers in that awful middle spot between on and off. They're constantly having what we call excessive daytime sleepiness. They're awake for a little while and then they can fall asleep very quickly and then they're awake again. So they're in this mixed no-man's land between wakefulness and sleep. It's an awful condition.

There is another terrible feature associated with narcolepsy called cataplexy. Remember when I described the paralysis that happens during REM sleep? Unfortunately that same mechanism of paralysis will kick in when narcoleptic patients are awake: they will just collapse down.

When you see it happening, it looks like they've had a sleep attack—which also happens—but that's not what's going on here. In this case they're awake, but they're paralyzed. And the trigger is strong emotion.

So if I'm a narcoleptic patient and you're behind the door and when I walk into the room you suddenly jump out and surprise me, that might be enough to paralyse me, and suddenly I'm on the floor.

HB: Is this related to what you were saying earlier about the amygdala when you were talking about emotional regulation?

MW: Exactly. We believe that what's happening is that this REM-sleep mechanism of paralysis becomes dysfunctional, but we're still starting to unpack the neural circuitry underlying this.

But imagine what it must be like—it's a truly terrible condition. I mean, think back to any moment in your life and try to imagine any decision that you've ever made that wasn't motivated by one of two, simple emotions: the need to stay away from something that was bad, or the need to achieve something that was going to be good.

Those two principles guide almost every single decision and action across our entire lives. Emotions are good. We like to engage in emotions. We like to try and stay away from the bad stuff and go after the good stuff. We're always self-medicating our emotions.

And now imagine trying to live a life with a condition where you are constantly reinforced **not** to experience strong emotions. You can't go to your daughter's party where there might be a clown and everybody's going to be excited because you'll start to get happy and then you'll just go into this paralysis attack. So you can't have a life with strong emotion. You have to stay away from it. You stay away from all the good stuff, the fuzzy stuff, the warm stuff. That's the really tragic aspect of narcolepsy. It's so sad.

HB: Does this also mean that they can't fully cleanse their "hippocampus USB stick" either, that they can't refresh their brains for learning as you were describing earlier?

MW: We're only just starting to ask those questions. Now that we're starting to understand the functions of sleep, we can go into disorders like narcolepsy, like insomnia, like depression, like PTSD, and try to make important progress.

We first begin by asking what it is, exactly, about the sleep of those particular people that is disrupted? Let's measure that—and for the most part, we now have a pretty good understanding of that for these different disorders.

So now we can go into our database of functions that are associated with those different stages of sleep, predict something specific that might be happening based on our current understanding of that condition, and test for that—see if it's actually present.

For example, you could imagine that one of the reasons that narcoleptic patients can suffer from cognitive or emotional impairments is because they're not getting the right amount of sleep, or sleep at the right time, to gain these benefits.

HB: That they're stuck in a vicious circle.

MW: Exactly. It might well be a spiral.

Questions for Discussion:

1. How does a deeper understanding of the particulars of a given condition, such as narcolepsy, increase one's tolerance towards those who suffer from it?

2. To what extent can examples of maladies and dysfunctions like narcolepsy often prove more helpful to our mechanical understanding than examining cases where systems function perfectly?

X. Lots To Do

Outstanding mysteries and public education

HB: Let's suppose that I'm an omniscient being and you can ask me three questions with regards to your research that you're desperate to know the answer to. What would those questions be?

MW: I would reserve one question for the unknown, because I think there's always probably something that I haven't thought of. So I'll add that to slot number three.

But I've got two key questions: The first is, *Why is sleep organized in the way it is?*

Earlier I described how you go into NREM sleep first and then, after about 70 or 80 minutes, you'll start to rise up and you'll have a short REM sleep period that will complete your 90 minutes. Then back down you go again into NREM sleep and the cycle repeats.

What's interesting, though, is that, while that 90-minute cycle remains stable across the night, the ratio of NREM to REM within those 90-minute cycles changes across the night.

So in the first half of the night, the majority of those 90-minute cycles are comprised of a lot of deep, NREM sleep and very little REM sleep. And as you push through to the second half of the night, that changes. The majority of those 90-minute cycles are comprised of REM sleep.

There's a remarkably odd, non-linear, but very deliberate, architecture of sleep that we can see in most species and in every individual each night.

So far sleep research has done a very good job of taking apart each of those stages of sleep and understanding each of their individual functions—

HB: But not integrating them all together.

MW: Exactly. We're suffering a horrible "hemineglect" in the sleep field. Nobody yet has a good idea of why sleep is structured in this way: *what is the benefit of this holistic, remarkably complex, structure of sleep?*

Personally, I think the answer is lurking within the depths of this sleep structure complexity, but right now, not only do we not have the answer, we don't even know how to ask the questions.

So I would love to know the answer to that.

My second question would be, *Is there a functional reason for dreaming above and beyond REM sleep?*

Right now those two things are sort of fused together in such a way that it's very difficult to get an experimental scalpel to separate them out.

If you think about a light bulb, for example, the reason we created a light bulb is to produce light. It turns out, however, that when you produce light in that way, you also produce something called heat. That's not the reason the light bulb was designed; that's just what happens when you produce light in that way.

It's what we call an epiphenomenon. That could also be the case with dreaming. Dreaming may have no function whatsoever. It might be solely an epiphenomenon. REM sleep may have functionality—it may be designed to help us in terms of emotional processing—but this conscious spin-off that we call "dreaming" might have no function whatsoever. Or it may have remarkable functional benefits.

There's a little bit of evidence out there right now that suggests that it's not *just* dreaming that's important, but perhaps **what** you dream: the content of your dreams may actually have a functional benefit.

The evidence right now is remarkably sparse and weak, so if I could have another question answered, it would be, *Is there an additional benefit of dreaming above and beyond simply the physiology on which it is based, which is REM sleep?*

HB: Very good. Here's another, somewhat different question that occurs to me as I'm listening to you: Do you have any personal meta-sleep issues related to the fact that you know so much about sleep? Do you lie awake at night unable to sleep stressing about how important it is that you fall asleep?

MW: It's funny; I've essentially become the Woody Allen neurotic of the sleep world. Normally my sleep is pretty good. I do practice what I preach. But let's say I've flown back home to England and I'm suffering from jet lag—

HB: It gets worse as you get older, by the way.

MW: Thanks for that. You're all heart.

So I'll be lying there in bed at night unable to fall asleep, and I'll start thinking, Okay, so my dorsolateral prefrontal cortex isn't shutting down. My noradrenalin and serotonin aren't starting to damp down, nor is my acetylcholine—and at that point of analysis, you're dead in the water for the next two hours. It's ironic on so many levels.

The other irony of sleep research is that, to understand what you're studying, you often have to deprive yourself of that very thing: you have to stay awake at night.

Luckily, I have wonderful students who work with me and they do all of that stuff so I get to have good nights of sleep, but there was a time when I was taking away the very thing that I was trying to understand the benefit of.

HB: How are you looked at by your colleagues? Is it weird to be looked at as the "sleep guy", or is everybody in a little box anyway?

MW: I would certainly say that if you're in sleep science, you're a novelty—I was going to say "oddity," but that implies negativity. Sleep science has often been considered almost a charlatan science. It's only been in the past 30 or 40 years that it's become a valid scientific field of study.

But dreams, they're still in the doldrums. To say that you're going to study dreams is academic suicide. You best have tenure before you start to study dreams.

HB: It's like saying that you want to study the foundations of quantum theory, I suppose.

MW: It's in that realm. Dreams are so ephemeral, so difficult to quantify. There is good hard, empirical science now being done on dreaming, but it's taken a while because of the terrible shadow of Freud.

The guy was very smart and he created a theory, but the problem with the theory was that it was non-scientific: you couldn't falsify it, but you couldn't prove it correct either. That was both its genius and its simultaneous downfall. Genius because you could never prove it wrong—and that's the reason why, more than a hundred years later, it's still with us. But its downfall is that it was never embraced scientifically because it's not an empirically testable hypothesis.

HB: I've talked to many people now in the field of cognitive science, broadly defined. It seems to me that the field has changed remarkably since fMRI and other real-time imaging technologies came into play, and since a lot of these scientific diagnostic practices were embraced.

And my sense, too, is that this has created a bit of a culture gap. On one hand you have the people who remain outside of all of that, and then you have those who have embraced all these modern technologies very rigorously.

Has that been your experience as well? Do you sometimes encounter old-style psychologists who tell you things like, "*Stay away from my Freud. Who the heck are you to be talking about dreams?! You don't understand anything. We understand the human condition and you're just measuring things with a bunch of knobs and dials*"?

Are there others who say, "*Enough with you guys. You haven't gotten us anywhere. You're just talking. Let's do some real science and push the boundaries*"?

These are very rough caricatures, I appreciate, but do you see those sorts of distinctions within the field? Is that a fair way of looking at things?

MW: I think within the field of sleep science, senior or junior, young or old, people are very much embracing modern technologies and understanding that that's going to pave the way to future discoveries.

I think when you stack up sleep science together with the world of psychoanalysis, there may be a gap between those two that may not necessarily allow those two worlds to sit happily together.

There are aspects of psychoanalysis that are non-empirical, non-scientific. That's not to say that it may not have a benefit, but it simply doesn't seem to be grounded in the empirical validity and replicability of science.

If there is a division, I think it's between the world of the sleep scientists and the world of the psychoanalysts. Those two things have stayed apart and I think they probably will remain apart.

HB: OK, but my question wasn't so much aimed at sleep science per se, but at a more general phenomenon. You're in the Psychology Department at Berkeley. I used to be an academic administrator, so I can imagine a situation where there's an opportunity for a faculty hire in Psychology and you hear from people who say something like, "*All of our retirees' posts are being filled by these young guys who are doing all this fMRI stuff. Enough of that. We have to have balance in the department.*" Does that sort of thing happen—let's not talk about at Berkeley in particular, but generally, throughout the field?

MW: Whenever new technologies, new theories, or new ways of measuring a specific process, are introduced, there is always speculation and resistance from those who won't adopt it, as well as positive claims from those who have rapidly adopted it.

Is that the case in my own field of sleep research? Yes, to a degree. But because we are such an odd little faction within the world of science, I think we're much more harmonious in our views. Now, that's not to say that at the end of a conference we all hold hands and sing "Kumbaya". There are divisions and debates, of course, but I think there's a remarkable degree of unification over adopting the new opportunities and agreement about where it will take us. And many of the old are adopting the new.

HB: Is the technology getting a lot better? Are there new things coming online? You do a lot of fMRI, but you also do a lot of EEG as well, right?

MW: Yes. We try not to be limited by the questions we can only answer with the technologies that are available, so we will rapidly develop technologies, or learn how to use new technologies, if we think they will help us answer the question. I'm not necessarily interested in technology for the sake of the technology. I'm always interested in questions, even more so than I am in answers, in fact. I think that's why people become scientists.

There are certainly new technologies that are starting to emerge within the animal research field. Optogenetics is a huge area. You can actually genetically engineer cells to be light sensitive, and you can use these to switch on some circuits within the brain, or switch them off, in very remarkably selective ways. That technology is now starting to come into the sleep world, and we're starting to be able to turn different sleep circuits on and off—and, as a consequence, amplify or impair the benefits that sleep gives.

I also think that our understanding of the brain mechanisms related to sleep and emotional regulation will, within the next 10 years, provide some significant insights into modern-day psychiatry.

I think one of the greatest areas of failing of modern medicine is the domain of psychiatry. In some ways, we're still where we were 20, 30, even 40 years ago. We don't have good drug treatments for depression or schizophrenia. NIMH, the National Institute of Mental Health, has really made a strong push over the last 5–10 years to change what they're funding to encourage *real* differences, because the field has, they believe, stagnated. And I support that.

I think sleep has a remarkably loud, potent, and causal story to tell the field of psychiatry. And I believe that in the next 5–10 years we'll have the evidence that really allows us to make those claims.

HB: Are you guys making headway in more general terms, in terms of the reputation of sleep science? At the beginning of this conversation you mentioned that sleep, along with eating, drinking and sex, serves an essential biological function. Are appreciable numbers of people starting to recognize this and paying more attention to your research?

MW: I think that, within the field of science and medicine generally, the recognition that sleep is an essential biological function has been embraced. As I said earlier, I think that there has been a switch, an inversion to now asking, "*Is there any tissue within the body, is there any mechanism within the brain that **isn't** benefited by sleep?*" So within the general area of science and medicine that recognition is definitely happening.

But what's still missing is the translation of that science to the general public. I would argue that, when it comes to the general public, we're at the same stage with sleep as we were with smoking 50 years ago.

If you roll the clock back 50 years or watch an episode of *Mad Men*, not only are people ignoring the biological risks of smoking, but if anything, society is *promoting* it.

Nowadays, you can't advertise smoking on television. We realized very quickly, from a public health standpoint, that it's so bad that we should take those ads away.

Yet you see advertisements everywhere for Red Bull and other kinds of "energy drinks" that are laced with alerting chemicals. And you hear people say things like, "*I'll sleep when I'm dead*"—which is definitely a very odd notion, to say the least.

So from a public health standpoint, the message still hasn't pervaded the public consciousness, but it's slowly getting there. That's why I think the responsibility lies with people like me to talk to people like you, in formats like this, and start trying to inform people.

I think we have a real responsibility, because it's the general public who is funding my work: it's the taxpayers' dollars that fund it. So not only do I have an ethical and moral responsibility to convey this message to the public, but I believe they deserve something for their money too. I think, on all of those levels, we need to get the message out there. Hopefully, by the grace of folks like you coming to speak to people like me, that will change. I'm a big advocate of trying to convey science to the general public and this type of medium is splendid. I feel very fortunate that you've asked me to do this.

HB: Well, it's been a delight for me. Thanks very much, Matt.

MW: Thank you. Sleep well.

Questions for Discussion:

1. Should governments regulate the sale and promotion of "energy drinks" just like they do for cigarettes and other tobacco products?

2. Do you think that overall Freud has had a positive or negative impact on our understanding of the significance of dreams?

3. Is all science necessarily falsifiable? (Readers with a strong interest in this topic are referred to Chapter 8 of the Ideas Roadshow conversation **Science and Pseudoscience** *with Princeton University historian of science Michael Gordin.)*

4. How has this conversation influenced your views about the importance of sleep?

Continuing the Conversation

Readers are encouraged to read Matthew's book, *Why We Sleep: Unlocking the Power of Sleep and Dreams*, which was published after this conversation and provides considerable additional detail about many of the issues discussed here.

Ideas Roadshow Collections

Each Ideas Roadshow collection offers 5 separate expert conversations presented in an accessible and engaging format.

- *Conversations About Anthropology & Sociology*
- *Conversations About Astrophysics & Cosmology*
- *Conversations About Biology*
- *Conversations About History, Volume 1*
- *Conversations About History, Volume 2*
- *Conversations About History, Volume 3*
- *Conversations About Language & Culture*
- *Conversations About Law*
- *Conversations About Neuroscience*
- *Conversations About Philosophy, Volume 1*
- *Conversations About Philosophy, Volume 2*
- *Conversations About Physics, Volume 1*
- *Conversations About Physics, Volume 2*
- *Conversations About Politics*
- *Conversations About Psychology, Volume 1*
- *Conversations About Psychology, Volume 2*
- *Conversations About Religion*
- *Conversations About Social Psychology*
- *Conversations About The Environment*
- *Conversations About The History of Ideas*

All collections are available as both eBook and paperback.